Biometrics in Identity Management

Concepts to Applications

For a listing of recent titles in the
Artech House Information Sercuity and Privacy Series,
turn to the back of this book.

Biometrics in Identity Management

Concepts to Applications

Shimon K. Modi

ARTECH HOUSE
BOSTON | LONDON
artechhouse.com

Library of Congress Cataloging-in-Publication Data
A catalog record for this book is available from the U.S. Library of Congress.

British Library Cataloguing in Publication Data
A catalog record for this book is available from the British Library.

ISBN-13: 978-1-60807-017-6

Cover design by Vicki Kane

© 2011 Artech House
685 Canton Street
Norwood, MA 02062

All rights reserved. Printed and bound in the United States of America. No part of this book may be reproduced or utilized in any form or by any means, electronic or mechanical, including photocopying, recording, or by any information storage and retrieval system, without permission in writing from the publisher.

All terms mentioned in this book that are known to be trademarks or service marks have been appropriately capitalized. Artech House cannot attest to the accuracy of this information. Use of a term in this book should not be regarded as affecting the validity of any trademark or service mark.

10 9 8 7 6 5 4 3 2 1

Contents

CHAPTER 1

Introduction 1

1.1	Basics of Biometrics	2
1.2	Types of Biometric Technologies	3
1.3	Biometric System	7
1.4	Biometric System Processes	8
1.5	Biometric System Architecture	10
1.6	Applications of Biometric Technologies	12
1.7	Classification of Biometric Applications	13
1.8	Summary	15
	References	15

CHAPTER 2

Fundamentals of Technical Evaluations 17

2.1	System Process Transactions	17
2.2	Types of Errors	18
	2.2.1 Acquisition Errors	18
	2.2.2 Matching Errors	19
2.3	Performance Metrics	19
	2.3.1 Failure to Enroll Rate (FTE)	19
	2.3.2 Failure to Acquire Rate (FTA)	19
	2.3.3 False Nonmatch Rate (FNMR)	20
	2.3.4 False Match Rate (FMR)	20
	2.3.5 Verification Performance Metrics	20
	2.3.6 Identification Performance Metrics	21
2.4	Type I and Type II Errors	22
2.5	Performance Curves	23
2.6	User-Specific Performance: Zoo Analysis	25
2.7	Evaluation Methodologies	27

	2.7.1	Technology Evaluation	28
	2.7.2	Scenario Evaluation	28
	2.7.3	Operational Evaluation	28
2.8	Design of Evaluation		29
	2.8.1	Sample Quality	30
2.9	Reporting Biometric Evaluations		31
2.10	Summary		31
	References		32

CHAPTER 3

Fingerprint Recognition — 33

3.1	Fingerprint Anatomy		33
3.2	History		34
	3.3.1	Henry Classification	35
	3.2.2	Automated Fingerprint Recognition	35
3.3	Fingerprint Presentation and Acquisition		36
	3.3.1	Inked Capture	36
	3.3.2	Latent Fingerprints	37
	3.3.3	Optical Sensors	37
	3.3.4	Solid State Capacitive Sensors	37
	3.3.5	RF Sensor	38
	3.3.6	Thermal Fingerprint Sensor	38
	3.3.7	Electro-Optical Sensor	39
	3.3.8	Multispectral Imaging Sensor	39
	3.3.9	Other Technologies	39
	3.3.10	Limitations of Acquisition Technologies	39
	3.3.11	Fingerprint Impressions	40
3.4	Fingerprint Feature Extraction		41
	3.4.1	Quality Assessment	42
3.5	Fingerprint Compression		43
3.6	Fingerprint Feature Matching		44
3.7	Classification and Indexing		45
3.8	Synthetic Fingerprint Generation		46
3.9	Automated Fingerprint Identification System (AFIS)		46
3.10	Standards		47
3.11	Evaluations		48
	3.11.1	Fingerprint Vendor Technology Evaluation (FpVTE)	48
	3.11.2	Minutiae Interoperability Exchange (MINEX)	49
	3.11.3	Minutiae Template Interoperability Test (MTIT)	50
	3.11.4	Fingerprint Verification Competition (FVC)	50
	3.11.5	Slap Fingerprint Segmentation Evaluations	51
3.12	Applications and Trends		52
	3.12.1	Integrated Automated Fingerprint Identification System (IAFIS)	52

	3.12.2 Personal Identity Verification (PIV)	52
	3.12.3 US-VISIT	52
3.13	Design and Deployment Considerations	53
	3.13.1 User Interaction	53
	3.13.2 Environment Issues	53
	3.13.3 User Demographics	54
	3.13.4 Interoperability	54
	3.13.5 Spoofing Attacks	54
	3.13.6 Large-Scale Systems	55
3.14	Summary	55
	References	55

CHAPTER 4
Face Recognition 59

4.1	History	59
4.2	2-D Face Recognition	60
	4.2.1 2-D Face Image Acquisition, Detection, and Segmentation	60
	4.2.2 Quality Assessment	61
	4.2.3 2-D Face Feature Extraction and Matching	62
	4.2.4 Video Surveillance	64
	4.2.5 Beyond the Visible Spectrum	64
4.3	3-D Face Recognition	65
	4.3.1 Face Acquisition	65
	4.3.2 3-D Face Modeling and Matching	66
4.4	Standards	66
4.5	Evaluations	68
	4.5.1 Facial Recognition Technology (FERET) Evaluations	68
	4.5.2 Face Recognition Vendor Test (FRVT)	69
	4.5.3 Face Recognition Grand Challenge (FRGC)	70
	4.5.4 Multiple Biometric Grand Challenge (MBGC) and Multiple Biometric Evaluation (MBE)	70
	4.5.5 Human-Machine Evaluation	71
	4.5.6 Other Evaluations	71
4.6	Applications and Trends	72
	4.6.1 Law Enforcement	72
	4.6.2 E-Passport	73
	4.6.3 U.S. Department of State Visa Application System	73
	4.6.4 Surveillance Applications	73
	4.6.5 Logical Access	73
	4.6.6 Airport Applications	73
	4.6.7 Personal Applications	74
4.7	Design and Deployment Considerations	74
	4.7.1 Pose and Expression	74
	4.7.2 Physiological Factors	74

	4.7.3	Environmental Factors	75
	4.7.4	Image Compression and Interoperability	75
	4.7.5	Subject Camera Distance and Motion	76
	4.7.6	Spoofing Attacks	76
4.8	Summary		76
	References		77

CHAPTER 5
Iris Recognition 79

5.1	Anatomy of Iris		79
5.2	History		80
5.3	Iris Image Acquisition		80
5.4	Feature Extraction		82
	5.4.1	Quality Assessment	83
5.5	Iris Feature Matching		84
5.6	Standards		85
5.7	Iris Capture at a Distance		85
5.8	Evaluations		86
	5.8.1	Independent Testing of Iris Recognition Technology (ITIRT)	86
	5.8.2	Iris Challenge Evaluation (ICE)	87
	5.8.3	IRIS06	88
	5.8.4	NIST IREX	89
5.9	Applications and Trends		90
	5.9.1	Privium System at Schipol Airport, Amsterdam	90
	5.9.2	U.K. IRIS Program	91
	5.9.3	U.S. Department of Defense (DoD) Iris Recognition	91
	5.9.4	United Arab Emirates (UAE) Expellee Program	91
5.10	Design and Deployment Considerations		92
	5.10.1	User Interaction	92
	5.10.2	Medical Conditions	92
	5.10.3	Nonideal Iris Images	93
	5.10.4	Environmental Effects	93
	5.10.5	Interoperability	93
	5.10.6	Spoofing Attacks	94
5.11	Summary		94
	References		94

CHAPTER 6
Hand Geometry Recognition 97

6.1	History	97
6.2	Image Acquisition	98
6.3	Feature Extraction	99

6.4	Feature Matching	101
6.5	Performance Evaluations	102
6.6	Standards	102
6.7	Applications and Trends	103
6.8	Design and Deployment Considerations	104
	6.8.1 Environmental Issues	104
	6.8.2 User Interaction	104
	6.8.3 Other Performance Considerations	105
6.9	Summary	105
	References	106

CHAPTER 7
Speaker Recognition — 107

7.1	Generating Voice	107
7.2	History	108
7.3	Speaker Recognition Systems	108
	7.3.1 Text-Dependent	109
	7.3.2 Text-Prompted	109
	7.3.3 Text-Independent	109
7.4	Information Levels	110
	7.4.1 Idiolectal	110
	7.4.2 Phonotactics	110
	7.4.3 Prosody	110
	7.4.4 Spectral Characteristics	111
7.5	Feature Extraction	111
	7.5.1 Signal Enhancement	111
	7.5.2 Mel Frequency Cepstral Coefficients (MFCC)	112
	7.5.3 LPC Cepstral Parameters	112
7.6	Feature Matching	112
	7.6.1 Distance-Based Methods	113
	7.6.2 Model-Based Methods	113
	7.6.3 Other Methods	114
7.7	Standards	114
7.8	Evaluations	115
7.9	Applications and Trends	117
7.10	Design and Deployment Considerations	119
	7.10.1 Voice Variations	119
	7.10.2 Background Noise	119
	7.10.3 Channel Effect	120
	7.10.4 Microphone Effect	120
	7.10.5 Duration of Samples	120
	7.10.6 Recording Attack	121

	7.11	Summary	121
		References	121

CHAPTER 8
Vascular Pattern Recognition — 123

	8.1	History	124
	8.2	Vein Pattern Acquisition	124
	8.3	Feature Extraction	125
	8.4	Feature Matching	126
	8.5	Facial Vascular Patterns	126
	8.6	Commercially Available Technologies	126
	8.7	Standards	127
	8.8	Performance Evaluations	128
	8.9	Applications and Trends	129
	8.10	Design and Deployment Considerations	130
	8.11	Summary	130
		References	131

CHAPTER 9
Dynamic Signature Verification — 133

	9.1	History		134
	9.2	Types of Signature Verification Systems		134
	9.3	Data Acquisition		134
	9.4	Feature Representation		135
	9.5	Feature Matching		136
	9.6	Standards		137
	9.7	Evaluations		138
	9.8	Trends and Applications		139
	9.9	Design and Deployment Considerations		139
		9.9.1	Inherent Variability	139
		9.9.2	Interoperability	140
		9.9.3	User Demographics	140
		9.9.4	Zero Effort Imposter Attempts	140
	9.10	Summary		141
		References		141

CHAPTER 10
Keystroke Dynamics, Retina, DNA, and Gait Recognition — 143

	10.1	Keystroke Dynamics		143
		10.1.1	History	144

		10.1.2	Feature Extraction and Matching	144
		10.1.3	Keystroke Dynamics Systems	145
		10.1.4	Standards	146
		10.1.5	Evaluations	146
		10.1.6	Trends and Applications	147
		10.1.7	Design and Deployment Considerations	147
	10.2	Retina Recognition		148
	10.3	DNA Recognition		149
	10.4	Gait Recognition		150
	10.5	Summary		151
		References		151

CHAPTER 11

Multibiometric Systems — 153

11.1	The Need for Multibiometric Systems		153
11.2	Multibiometric System Design		154
	11.2.1	Multimodal	155
	11.2.2	Multi-Instance	155
	11.2.3	Multisensor	155
	11.2.4	Multialgorithmic	155
	11.2.5	Multisample	155
11.3	Data Acquisition		156
11.4	Levels of Fusion		156
	11.4.1	Sensor Level Fusion	156
	11.4.2	Feature Level Fusion	157
	11.4.3	Score Level Fusion	157
	11.4.4	Rank Level Fusion	161
	11.4.5	Decision Level Fusion	161
	11.4.6	Quality-Based Fusion	162
11.5	Standards		162
11.6	Evaluations		163
11.7	Trends and Applications		165
11.8	Design and Deployment Considerations		166
	11.8.1	Cost	166
	11.8.2	Correlation	166
	11.8.3	Human Factors	166
	11.8.4	Fusion Architecture	167
	11.8.5	Quality of Samples	167
	11.8.6	Zoo Analysis	167
	11.8.7	Privacy	167
11.9	Summary		168
	References		168

CHAPTER 12

Biometric Standards — 171

12.1 The Importance of Standards — 171
12.2 Standards Development Organizations (SDO) — 172
    12.2.1 ISO/IEC — 172
    12.2.2 National Standards Bodies — 173
    12.2.3 BioAPI Consortium — 174
    12.2.4 NIST — 174
    12.2.5 OASIS — 174
    12.2.6 International Telecommunication Union (ITU-T) — 175
12.3 Working Groups — 175
    12.3.1 WG1 Harmonized Biometric Vocabulary — 175
    12.3.2 WG2 Biometric Technical Interfaces — 176
    12.3.3 WG3 — 177
    12.3.4 WG4 — 178
    12.3.5 WG5 — 179
    12.3.6 WG6 — 179
    12.3.7 Sample Quality Standards — 180
    12.3.8 Conformance Testing — 181
    12.3.9 Security Standards — 182
12.4 Standards Used in Law Enforcement — 183
    12.4.1 ANSI/NIST-ITL — 183
    12.4.2 FBI EBTS — 183
    12.4.3 DoD EBTS — 183
    12.4.4 INTERPOL INT-I — 184
12.5 Adoption of Standards — 184
12.6 Summary — 184
    References — 185

CHAPTER 13

Biometric Testing and Evaluation Programs — 187

13.1 Biometric Testing: Why Is It Required? — 187
13.2 Biometric Data Considerations — 188
13.3 Unimodal Performance Evaluations and Research Databases — 188
    13.3.1 Fingerprint Recognition — 189
    13.3.2 Face Recognition — 191
    13.3.3 Iris Recognition — 195
    13.3.4 Speaker Verification — 197
    13.3.5 Signature Verification — 200
    13.3.6 Ocular Recognition — 200
13.4 Multibiometric Evaluations and Research Databases — 201
13.5 Comparative Tests — 204

13.6	Liveness Detection Evaluations	205
13.7	Summary	205
	References	206

CHAPTER 14

Desiging and Deploying Biometric Systems — 209

14.1	Implementation Plan	209
14.2	Application Scope	210
14.3	Technology Selection	210
14.4	User-System Interaction	213
	14.4.1 Usability Principles	213
	14.4.2 Usability Design	215
14.5	Operational Processes	216
14.6	Privacy Principles	216
14.7	Architecture	217
14.8	Application Development	217
	14.8.1 Application User Interface	218
14.9	Policy Development	218
14.10	Design Validation	219
14.11	Disaster Recovery Plan	219
14.12	Maintenance	220
14.13	Summary	221
	References	221

CHAPTER 15

Biometric System Security — 223

15.1	System Security Analysis	223
	15.1.1 Subsystem Vulnerabilities	224
	15.1.2 Transmission Vulnerabilities	225
	15.1.3 Process Vulnerabilities	226
15.2	Spoofing and Mimicry Attacks	227
	15.2.1 Fingerprint Recognition	227
	15.2.2 Face Recognition	228
	15.2.3 Iris Recognition	228
	15.2.4 Hand Recognition	229
	15.2.5 Speaker Recognition	229
	15.2.6 Vascular Pattern Recognition	230
	15.2.7 Mimicry Attacks	230
	15.2.8 LiveDet Competition	231
	15.2.9 TABULA RASA	231
15.3	Standards	232
15.4	Synthetic Biometric Samples	232

	15.4.1 Synthesis Approaches	233
15.5	Summary	234
	References	235

CHAPTER 16
Privacy Concerns in Biometric Applications — 237

16.1	Privacy Invasive Technologies (PIT)	237
	16.1.1 Permanent Identifier	238
	16.1.2 Function Creep	238
	16.1.3 Tracking and Profiling	238
	16.1.4 Data Breach	238
	16.1.5 Data Misuse	238
	16.1.6 Unauthorized Collection	239
16.2	Privacy-Enhancing Technologies (PET)	239
16.3	Privacy Frameworks	240
16.4	Privacy Impact Assessment (PIA)	242
16.5	Standards	244
16.6	Privacy-Enhancing Design Principles	246
16.7	Future Challenges	246
16.8	Summary	247
	References	247

Acronyms	249
Glossary	251
About the Author	257
Index	259

CHAPTER 1

Introduction

In today's digital infrastructure we have to interact with an increasing number of systems, both in the physical and the virtual worlds. Identity management (IdM), the process of identifying individuals and controlling access to resources based on their associated privileges, is becoming progressively complex. Today IdM has become an inescapable fact of life, from logging onto e-mail accounts and accessing corporate networks to boarding a flight. This has brought the spotlight on the importance of an effective and efficient means of ascertaining an individual's identity. Traditional recognition techniques are based on *something that you know* (i.e., passwords) or *something that you possess* (i.e., tokens and ID cards). Completing the triad of recognition technologies, shown in Figure 1.1, is *biometrics*, which is defined as the automated recognition of humans based on biological or behavioral characteristics [1]. Although the use of biometric technologies, such as fingerprint recognition, face recognition, and iris recognition, is a more recent phenomenon, the use of human features for recognition can be traced back to the fifth century B.C. Archeologists have discovered evidence that indicates that Babylonian and Chinese civilizations used fingerprints to associate earthen pots with their creators. Since the nineteenth century, fingerprints and their utility in recognition have been studied using scientific methods [2]. Biometric technologies have also made appearances in science fiction novels for over half a century—Isaac Asimov referred to the use of human characteristics for identification purposes in his book *Foundation and Empire,* published in 1952. In the last two decades there has been a rapid growth of biometric technologies in government, industry, and personal applications as the traditional means of recognition have come under increasing pressure to keep up with today's IdM demands. The always-connected, always-on nature of today's systems has made it easier for threats to launch attacks, which have led to the demand for strong authentication mechanisms.

Knowledge-based methods such as passwords or passphrases can be forgotten, stolen, or used surreptitiously. Possession-based methods such as tokens and ID cards are also prone to the same vulnerabilities, with the additional possibility of misplacing them. These vulnerabilities can be exposed by a variety of security threats and expose the owners to financial and legal risk. There are administrative costs of reissuing the password or token, potential legal and regulatory fines, and an adverse impact on the owners' credibility. Nonrepudiation, which is the concept of an individual not having the ability to disown a particular action performed by

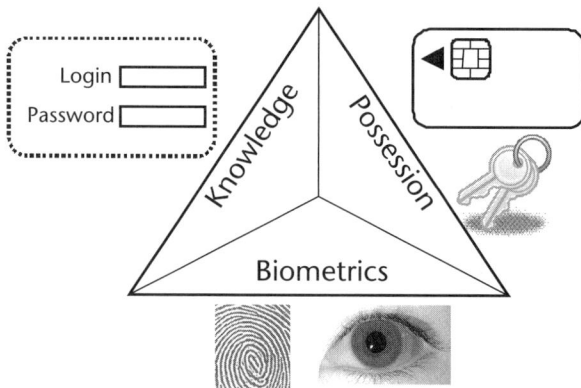

Figure 1.1 Types of recognition methodologies.

the individual, is impossible to determine for systems that use passwords or tokens. The physiological or behavioral characteristics used for biometric recognition cannot be forgotten or misplaced. Biometric technologies offer a reliable method of recognition in addition to providing nonrepudiation. Biometrics is being increasingly used in government programs such as border control and government-to-citizen services, along with consumer-facing applications in healthcare and finance sectors. Biometric technologies offer reliable and efficient recognition that is necessary as our real and virtual worlds are further enmeshed.

The expectations of what biometric technologies can achieve will grow with its increased adoption. Biometrics is not a silver bullet; it cannot provide 100% security, nor can it provide a reliable solution for every problem. As with knowledge- and possession-based methods, biometric technologies also have their weaknesses. The intention of this book is to serve as a guide to practitioners and applied researchers in the area of biometrics. Various commercially available biometric technologies are discussed in this book, but rather than focusing on the underlying algorithms, this book focuses on factors that drive the practical implementation of this technology and, in the process, seeks to serve as a bridge between researchers and practitioners. There are several excellent books that cover the underpinnings of biometric sensors and algorithms and are referenced when necessary. This book will focus on development and deployment issues facing biometric technologies and several open questions that need to be answered for increasing its adoption.

1.1 Basics of Biometrics

The etymology of the word *biometrics* can be found in the Greek words *bios*, which means life, and *metron*, which means measure. Biometric technologies are based on several different physiological and behavioral characteristics such as fingerprints, face structure, iris patterns, and voice signal. Although there are many human characteristics that are candidates for use in biometric technologies, they must satisfy the following criteria to be useful [3]:

- *Universality:* They should be present in the entire normal population.

- *Uniqueness:* They should be relatively unique and differentiated among every individual of the normal population.
- *Collectible:* They should be captured in real time without any intrusions on privacy.
- *Permanence:* They should stay relatively stable throughout the period of use of the particular characteristic.
- *Performance*: They should perform accurately and consistently in real time.

An implementation of a biometric technology has to take into account practical considerations so that it can be used effectively in an operational scenario. From a security perspective, the technology should provide a *liveness detection* capability to protect against spoofing attacks. Spoofing, which is the act of presenting a nonlive or fake biometric sample, can render useless the security effectiveness of a biometric technology. Liveness detection and antispoofing techniques are discussed in detail in Chapter 15. Other practical considerations include:

- *Throughput:* The system should provide a result in real time that does not inconvenience the user or impact the organization's processes.
- *Usability:* The system should be intuitive to use and provide a satisfactory experience to the user.
- *Scalability:* The system should be capable of handling an increasing amount of data without any significant impact on performance, throughput, and usability.
- *Acceptability:* The system should be sensitive to privacy and cultural concerns of the users.

Not all biometric technologies satisfy these requirements to the same degree. The specific application will drive the priority given to each of these criteria and the final selection of a particular biometric technology, but they all should be considered as part of the decision function. Table 1.1 summarizes the selection criteria for biometrics from three different perspectives.

1.2 Types of Biometric Technologies

Over a century of research in biometrics has led to development of recognition technologies based on several physiological and behavioral traits. Fingerprint

Table 1.1 Selection Criteria for Biometrics

Inherent Traits	*System Specific*	*User Specific*
Universality	Liveness detection	Usability
Uniqueness	Throughput	Acceptability
Collectible	Usability	
Permanence		
Performance		

recognition, face recognition, iris recognition, and voice recognition have a higher level of public awareness because of consumer-facing applications, media coverage, and movies, while there are others such as vein recognition and keystroke dynamics that are relatively unknown. Traditional biometrics literature categorizes technologies into two distinct classes based on how the trait being measured is generated. *Physiological* biometric technologies use anatomical features such as fingerprints, face, and iris structure. *Behavioral* biometric technologies use actions or mannerisms that are acquired or learned over time such as signature, gait, and typing pattern. The process of voice generation is affected by behavioral (intonation, accent) and physiological (vocal chords, nasal cavity, oral cavity) factors and so is considered to be a mixture of both. Generally, physical characteristics provide a more consistent reading as they are minimally affected by the behavior of the individual. They are also considered to be more accurate than behavioral biometrics, but research has shown that it is possible to effectively use behavioral biometrics to improve security and convenience. Table 1.2 lists a summary of existing biometric technologies.

From a system view biometric technologies can be categorized on a continuum where the extreme ends are behavioral and physiological (Figure 1.2), instead of two distinct categories. All biometric technologies require users to interact with a biometric sensor, which is impacted to a certain degree by the behavior of the individual. So although an underlying fingerprint recognition algorithm uses only physical characteristics, the capture process is impacted by how the user interacts with the fingerprint sensor.

The biometrics domain is an extremely dynamic one with several new technologies in the research pipeline, some of which will be commercialized in the near future. Keeping in mind the theme of this book, the well-established and commercialized biometric technologies are briefly discussed here.

Table 1.2 Description of Biometric Technologies

Recognition Technology	Type	Physical Interaction	Example Use Case
Fingerprint	Biological	Required	Network log-on, computer log-on, criminal identification
Face	Biological	Not required	Network log-on, computer log-on, criminal identification
Iris	Biological	Not required	Network log-on, border and immigration control
Hand geometry	Biological	Required	Time and attendance, door access
Voice	Biological and behavioral	Not required	Identity verification in mobile commerce and e-commerce
Vascular pattern	Biological	Required	Identity verification in healthcare
Dynamic signature verification	Behavioral	Required	Identity verification in credit card transactions
Keystroke dynamics	Behavioral	Required	Complement password authentication
DNA	Biological	Required	Law enforcement
Retina	Biological	Not required	Identity verification for physical access
Gait recognition	Behavioral	Not required	Surveillance applications
Ear recognition	Biological	Not required	Identity verification for physical access

1.2 Types of Biometric Technologies

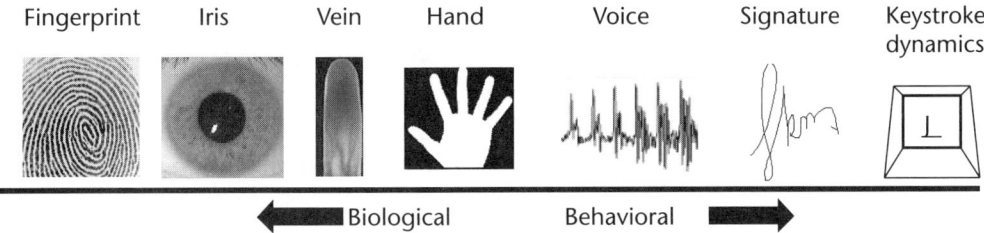

Figure 1.2 Examples of biometric traits.

- *Fingerprint recognition* uses the pattern found on the skin of fingers to identify individuals. This is the oldest and most widely adopted biometric technology and, as a result, is the most mature of all biometric technologies. The fingerprint is rich in detail and several different methods for capturing, processing, and comparing them have been tested successfully. Fingerprint recognition is discussed in detail in Chapter 3.

- *Face recognition* uses the structure and the spatial geometry of the features and landmarks such as the nose, eyes, lips, and jawline. This technology has made great improvements in the last two decades and can be performed on 2-D and 3-D images of the face. Face recognition is discussed in detail in Chapter 4.

- *Iris recognition* uses the pattern formed by muscle tissue and cell structures in the iris region of the eye. The iris is the circular ring surrounding the pupil and its main function is to control the size of the pupil and regulate the amount of light entering the eye. The iris image is captured using infrared illumination and a camera, and not using invasive lasers, as is often depicted in media reports and movies. Iris recognition has attracted a lot of commerical and research interest in the last decade, and is discussed in detail in Chapter 5.

- *Hand recognition* uses the contour of the hand, length and thickness of fingers, and spatial distance between other landmarks on the hand. This technology has been around since the 1970s and is heavily favored in physical access control applications. Hand recognition is discussed in detail in Chapter 6.

- *Voice recognition* uses the vocal characteristics such as pitch, intonation, and vocal speed. The voice of a person is affected by several acquired geocultural factors, as well as physiological factors such as the shape and size of the vocal chords, the nasal cavity, and the larynx. This technology has attracted a considerable amount of interest from intelligence and surveillance agencies, but has traditionally seen low adoption in the commercial sector. That is now changing with the proliferation of mobile devices because a standard microphone is suitable for capturing voice samples. Voice recognition is discussed in detail in Chapter 7.

- *Vascular pattern recognition systems* use vein patterns for recognition. Veins carry deoxygenated blood from the various parts of the body back to the heart and research has found the pattern formed from the vascular network is relatively unique and permanent. Commercial products capture vein patterns using infrared illumination, and currently they focus on the finger, the palm, and the back of the hand. Vascular pattern recognition is discussed in detail in Chapter 8.

- *Dynamic signature verification* uses features such as the velocity, direction, number of strokes, time of each stroke, and pressure applied by the user during the signing process. The legal tender associated with a signature makes it an appealing technology for use in identity verification applications. Dynamic signature verification is discussed in detail in Chapter 9.
- *Keystroke dynamics* uses the typing rhythm of a user on keyboards or other type of input devices for authentication. Most commercial systems use a standardized keyboard as the input device and do not require any specialized hardware. The accuracy and reliability of this technology are still improving and are currently considered a good complement to password-based systems in a multifactor authentication. Keystroke dynamics is discussed in detail in Chapter 10.
- *DNA identification* has typically been used in forensic sciences, but is now being pursued as a biometric technology. There are still technology issues such as invasive data capture and processing time, which runs in hours and not in seconds. The definition of biometrics specifically refers to *automated recognition*, and although current DNA analysis is not completely automated, future advances in this technology should not rule out such a possibility. Its high level of distinctiveness among individuals makes it an extremely promising technology for the future.

There are other biometric technologies such as retina recognition, gait recognition, ear lobe recognition, scent recognition, hand gesture recognition, knuckle recognition, and others that are being actively researched by the scientific community. These technologies are not discussed in-depth in this book, as they have not transitioned to real-world deployments. Interested readers are encouraged to read scientific journals and publications to gain a better understanding of them.

All of these technologies are unimodal, which means that they use a single trait for recognition. The ability to fuse multiple biometric technologies to enhance performance and eliminate weaknesses has led to the design of multibiometric systems. These systems combine multiple traits (e.g., finger and face), multiple units of the same characteristic (e.g., different fingers), or a variety of other information sources. Multibiometrics has received increasing interest as the limitation of unimodal systems in large-scale applications is becoming apparent. Multimodal biometrics is discussed in depth in Chapter 11.

There is no "best" biometric technology, and one of the goals of this book is to highlight the advantages and disadvantages of different technologies with respect to different scenarios. The selection criteria for biometric technologies discussed earlier is affected by a variety of factors such as the user population, deployment environment, and requirements of the application. On completing this book, readers will have the necessary tools to make an educated decision using a holistic approach.

1.3 Biometric System

A biometric system is essentially a pattern recognition engine that uses a representation of human traits as its input. A generalized biometric system can be viewed as a functional combination of five subsystems, as shown in Figure 1.3 [4]:

1. *Acquisition:* This subsystem is responsible for capturing the raw biometric sample from a user. Acquisition is typically performed using a sensor that could require physical interaction with the user. This is the only point of interaction between a user and the biometric system and hence the source of all interaction errors that are injected into the system. Errors introduced here will propagate through the rest of the system and increase the probability of system errors.
2. *Signal processing:* This subsystem is responsible for extracting features that represent the uniqueness of the sample. This module preprocesses the sample for enhancement, performs quality assessment, and creates a feature representation for subsequent use in either storage or matching. The quality assessment component is an extremely important part of this subsystem, as it determines if the sample is suitable for feature extraction or if it needs to be recaptured. This is a compact representation of the raw signal and is typically designed to be noninvertible (i.e., it is computationally impossible to recreate the raw sample from the feature representation).
3. *Data storage:* This subsystem stores the feature representation produced by the signal processing subsystem. This feature representation that is stored for future use is also called a *template* in the biometrics domain. Data storage can either be centralized (i.e., stored on a server) or localized (i.e., stored on a smart card or personal storage media).
4. *Matching:* This subsystem compares two feature representations and produces a similarity score. The similarity score is the degree of confidence that the two original samples are from the same individual. A biometric matching subsystem is *probabilistic* in nature—two samples from the same individual will never provide a perfect match. In comparison, password

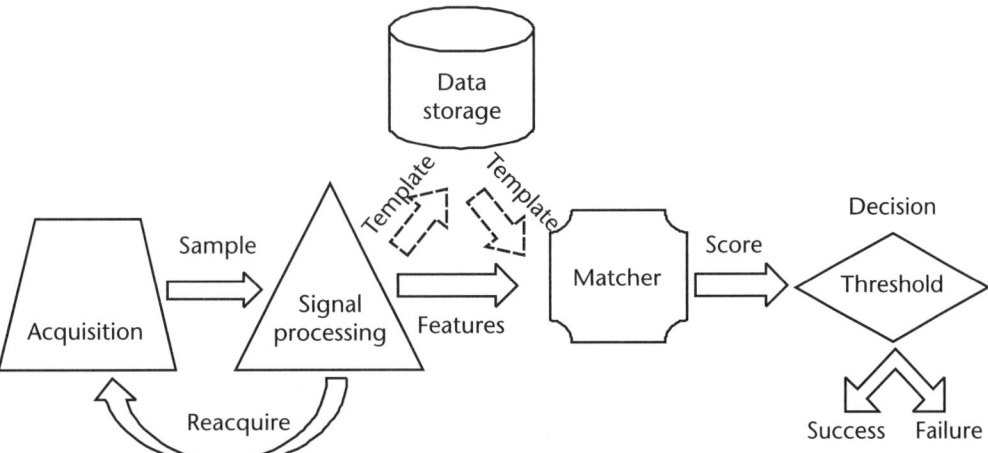

Figure 1.3 Biometric system model.

and cryptographic techniques require a perfect match in order to declare it successful. Due to human interaction with the acquisition subsystem, successive samples from the same individual are never exactly the same. Instead of providing a binary response, a similarity score is calculated.

5. *Decision making:* This subsystem uses the similarity score generated by the matching subsystem and compares it to a *threshold* value to generate a success or failure decision. The *threshold* value indicates the variability allowed between two biometric samples for them to be considered from the same source. The threshold value is also a reflection of the acceptable level of risk of the biometric system owner. The threshold value plays an instrumental role in decision errors produced by biometric systems, which are discussed in Chapter 2. The decision about the specific threshold value should be taken after careful discussions among all organizational units that are affected by the biometric system.

Biometric systems are required to handle variations in samples. A complete biometric system is, in essence, a pattern recognition machine with the dual goals of maximizing interclass variance and minimizing intraclass variance. Interclass variance is maximized by using features from subjects that are distinct between individuals. Intraclass variance is minimized by using features from subjects that remain stable over time and can be captured consistently. These dual principles of *distinctiveness* and *stability* form the underpinnings of any biometric system.

1.4 Biometric System Processes

A biometric system has to create a template from an individual's biometric features and compare subsequent samples to the registered template. In the context of a biometric system, these processes are separately called *enrollment*, *verification*, and *identification*, as illustrated in Figures 1.4, 1.5, and 1.6, respectively. During *enrollment* an individual provides his or her biometric sample to the system and a template is generated and stored for future use. During *verification* an individual makes a claim to an identity, for example, providing a user ID number along with a biometric sample. Knowledge-based and token-based recognition techniques always work in verification mode since they require additional information such as usernames or

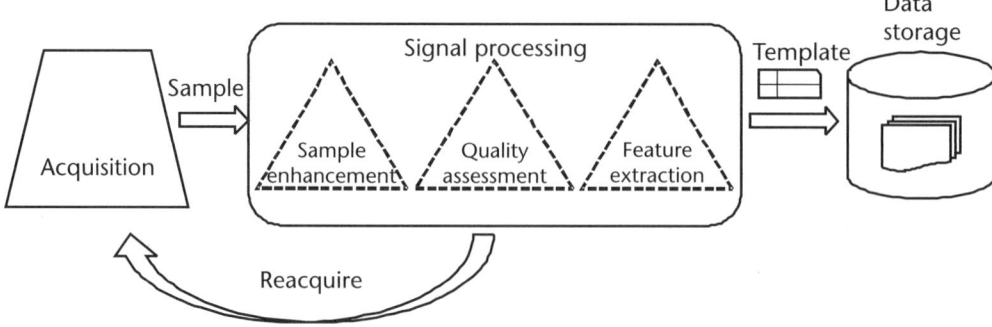

Figure 1.4 Enrollment process.

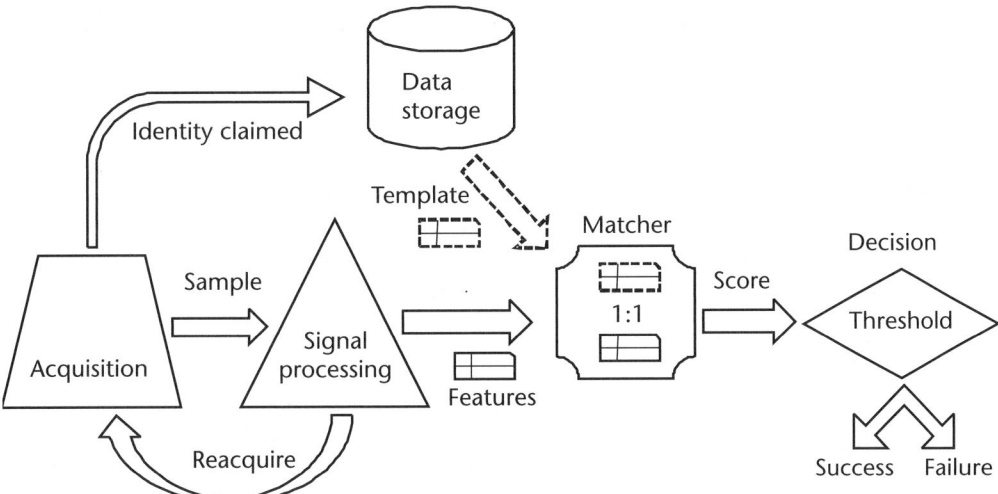

Figure 1.5 Verification process.

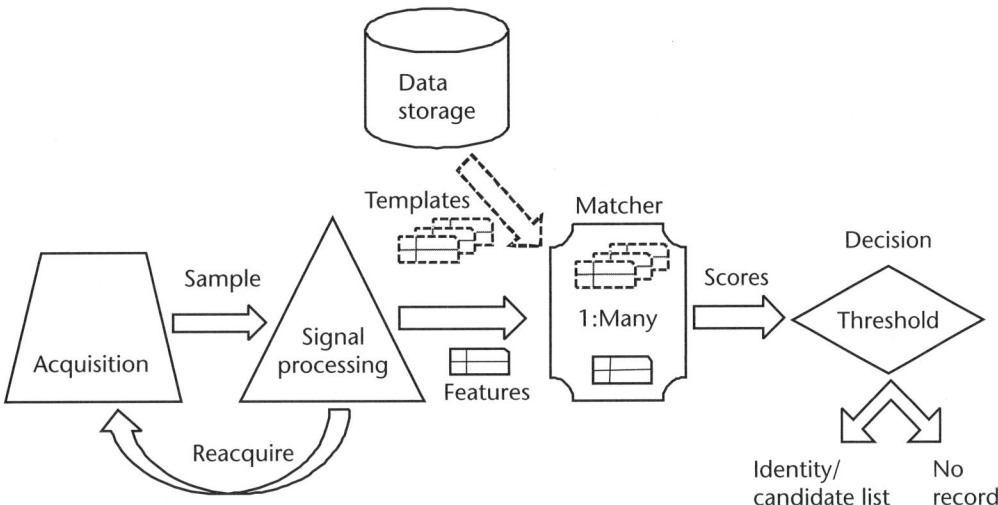

Figure 1.6 Identification process.

the physical token itself. In identification an individual does not make a claim to a specific identity, but rather lets the biometric system decide if he or she is a member of the enrolled list of individuals. In literature verification is often referred to as 1:1 matching and identification is referred to as 1:many matching. Certain biometric applications that operate in the identification mode might ask for a list of the closest matching candidates instead of only the best match. Such a list is called a *candidate list* and the maximum number of individuals on the list can be defined by the system administrator. Candidate lists are popular in applications that require human intervention such as law enforcement. Take, for example, the process of comparing an unknown fingerprint against a large database of known criminals, which returns a candidate list that is further examined by a human expert. A biometric identification process that does not require any human intervention is called a *lights-out process*. The eventual goal of all biometric systems is to operate in lights-out mode, but the consequences of mismatch errors will require human intervention for outcomes with legal repercussions.

Identification can be performed on either a *closed set* or an *open set*. Closed-set identification is a biometric task that determines if an input sample belongs to an individual who is a member of the system and is already known to the system. In closed-set identification the user who is to be identified has to be enrolled in the system. Open-set identification is a biometric task that determines if the input sample belongs to an individual who is a member of a system. The key difference between the two is that in open-set identification the user providing the input sample does not have to be enrolled in the system.

The identification process can be applied in two different ways: *positive* and *negative* identification. The basic process for both is the same; what differs is the interpretation of result. The goal of positive identification is to prove to the system that the user is known to the system, while the goal of negative identification is to prove to the system that the user is not known to the system. Positive identification is used for checking if a person is a member of an authorized list of individuals. Negative identification is used to check for multiple enrollments or if an individual is on a blacklist. For example, negative identification is being performed when an individual who is being issued a new driver's license is checked against all existing license holders to ensure that multiple licenses are not given to the same individual.

For a real-world system vetting user credentials is extremely important part of the enrollment process. During enrollment the biometric template is linked to an identity that is established based on veracity of credentials provided by the individual. An error during the enrollment process is perpetuated throughout the identity credential life cycle and defeats the purpose of strong authentication.

1.5 Biometric System Architecture

Any technology system should not be regarded as an island onto itself; today's enterprise infrastructure consists of several systems integrated to provide a seamless infrastructure. The possible locations for storage and matching systems provide system designers with the flexibility to create systems based on a variety of architectures, as shown in Figure 1.7. These locations can be categorized into the following centralized server, local workstation, peripheral device, and physical token.

The server is defined as a centrally located system that is at a different physical location than the requesting client and typically is connected to several clients. The

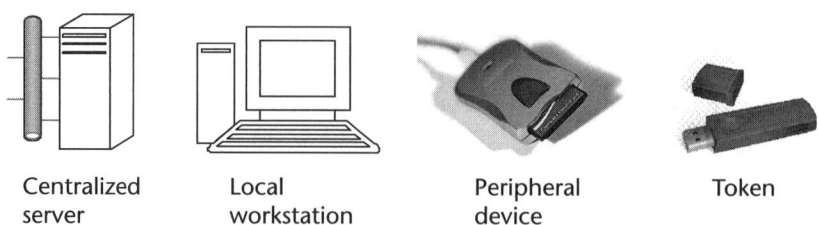

Figure 1.7 Possible locations for storage and matching.

1.5 Biometric System Architecture

local workstation is where a user initiates interaction with the biometric system. The peripheral device can also be connected with the local workstation using input/output ports or an embedded device. Physical tokens are smartcards, PCMCIA cards, and other small-scale devices that are capable of storing data or matching templates. The INCITS M1 Ad Hoc Group on Biometrics and E-Authentication (AHGBEA) published a report in 2007 that described different biometric architectures based on the location of the storage and matching subsystem [5]. There are 16 (4^2) configurations possible, illustrated in Figure 1.8, but not all of them are feasible in a practical implementation.

The architecture of all biometric system deployments will belong to one of these possible configurations. Several fingerprint recognition products are available that replace password authentication in an active directory with fingerprint recognition in enterprise network applications. These products, which are an example of distributed acquisition and centralized storage/matching architecture, allow organizations to centralize the IdM processes. The Seafarers' Identity Document (SID), which is issued by countries who are members of the International Labor Organization (ILO), contains the owner's fingerprint template on the card itself. For verification the seafarer provides his or her fingerprint sample to a kiosk along with the template stored on his or her card. In this scenario the card acts as the storage location and the matching occurs on the machine in the kiosk. This is an example of architecture that uses a physical token for storage and a local workstation for matching.

Along with technical feasibility, the final architecture is also driven by a mixture of user concerns and performance metrics required by the system administrators. The primary user concern is privacy of information and sharing of data with third parties. In addition, maintaining confidentiality and integrity of the biometric

		Matching			
		Server	Workstation	Peripheral	Token
Storage	Server	Law enforcement			
	Workstation		Computer access		
	Peripheral				
	Token	Personal verification			Match-on-Card

Figure 1.8 Biometric architectures and example applications.

data and preventing unauthorized individuals from accessing this data should be a priority for system administrators. Data protection laws also have to be followed due to the financial and reputational ramifications of data breach incidents. Chapter 14 discusses in detail the design consideration for solution architects and program managers.

1.6 Applications of Biometric Technologies

Biometric applications are seen as a mechanism to counter the risks of identity fraud and establish a strong link with the identity credentials. By removing the need to remember passwords or carry physical tokens, biometric technologies enhance user convenience and improve security. There are several candidate domains for applications of biometric technologies, and they are discussed here. Figure 1.9 illustrates the trade-off between accuracy and convenience for the various domains of applications.

1. *Government applications* account for the maximum number of biometric deployments. Among them, law enforcement has been the forerunner in adopting biometric technologies, with the first use dating back to 1892 for prisoner identification using fingerprints [6]. Biometrics is increasingly considered for government-to-government and government-to-citizen applications that require citizen identity management. Biometric national ID cards have been introduced in the Philippines and Malaysia, and Bangladesh has implemented voter identification based on biometrics. Several countries have biometrics-based IdM for border control, welfare disbursement, and other government-to-citizen services.
2. *Commercial applications* in banking, retail, healthcare, and other sectors have used biometrics for over a decade, but the adoption and usage rate have not kept up with the pace of government applications. While government applications can focus on accuracy at the expense of user convenience,

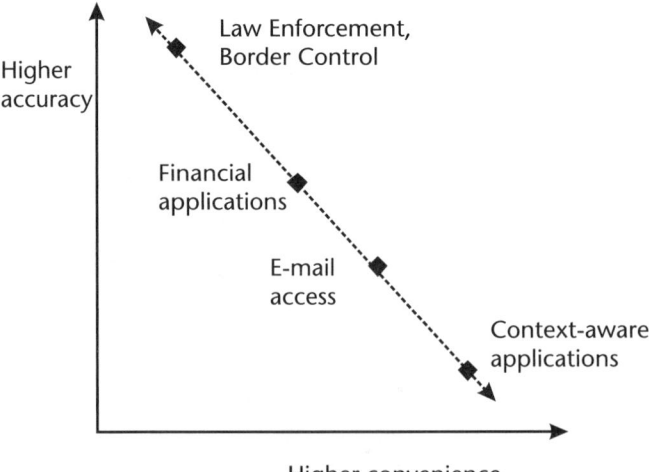

Figure 1.9 Security versus convenience.

commercial applications have to balance accuracy and user convenience. Adoption of biometrics in commercial applications will primarily be driven by the demands of meeting industry-specific regulations along with an increased focus on user experience and convenience.

3. *Forensic applications* are a natural extension of biometric technologies, as their underlying principles have a lot of overlap. Although this is not related to previous two categories in terms of security and convenience, forensic examinations such as dead body identification is possible using fingerprints, DNA, and other biometric traits.

4. *Personalization/context-aware applications* represent a new class of applications that can take advantage of biometric technologies. This class of applications is not designed for the purpose of access control, but instead for personalizing settings or configurations for using a particular device or service in a specific context. For example, a personalized application based on face recognition can be used in a car to recognize the driver and adjust the height of the seat and the steering wheel to the settings defined by the driver.

According to a market research report by the International Biometric Group (IBG), the market for biometric technologies is expected to grow from $3.4 billion in 2009 to $9.3 billion in 2014, as illustrated in Figure 1.10 [7]. It is expected that fingerprint recognition will contribute to a majority of the growth, along with face, iris, voice, and vein recognition as significant contributors as well. The need to reduce identity fraud, increase border security, and improve convenience for users will be the biggest drivers of growth.

1.7 Classification of Biometric Applications

An IdM can be classified along several dimensions based on the requirements of the application and the operational environment. These dimensions are also referred to as Wayman's Taxonomy [8]:

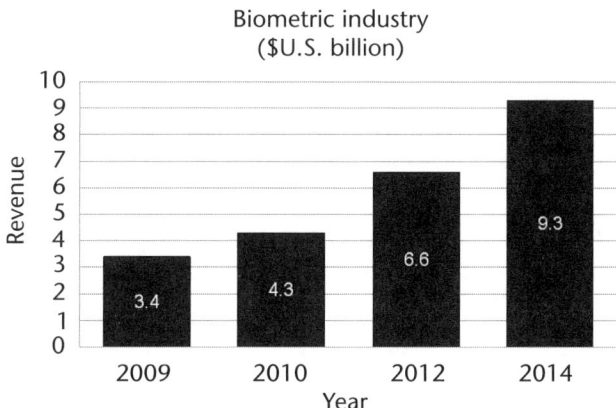

Figure 1.10 Biometric market growth ($ billion) [6].

1. *Overt or covert:* This refers to the user's awareness and approval for having his or her sample captured and processed by a biometric system. Although most biometric applications are overt, surveillance applications might operate in covert mode.
2. *Cooperative or noncooperative:* This refers to the behavior of an intruder who is interacting with a biometric system. The objective of the intruder is to circumvent the security procedure. Depending on positive or negative identification, the intruder will determine his or her specific behavior. In negative identification an intruder who is on the list does not want to be detected. In such a case it is in his or her best interest to be noncooperative and avoid detection. In a positive identification application the intruder wants to be positively identified, albeit as another individual. In this case it is in his or her best interest to cooperate and increase his or her probability of getting accepted.
3. *Habituated or nonhabituated:* This refers to how often a user interacts with a biometric system and level of training required for proper functioning of the system. Most systems will initially have to cater to nonhabituated users who will gradually become habituated with repetitive use.
4. *Supervised or nonsupervised:* Certain biometric systems have an operator or supervisor who oversees the system processes and intervenes if required. Best practices suggest that enrollment should be supervised for optimal results, and verification or identification should be nonsupervised based on application requirements. Law enforcement and border control applications are classic examples of supervised systems. Noncooperative systems should be operated in a supervised mode for it to be effective.
5. *Standardized or nonstandardized environment:* This refers to the consistency of the environment throughout the entire deployment. Biometric systems are affected by external factors such as background and illumination, and keeping them consistent is one way of improving performance of the system.
6. *Closed or open systems:* This refers to the requirement of the system to share data with other systems. Law enforcement applications are an example of an open system, whereas an enterprise log-on system is likely to be a closed system.
7. *Public or private:* This refers to the relationship between the user and the system owners. Government applications such as border control and welfare disbursements are examples of public systems, whereas network log-on for employee verification is an example of a private system. User concerns have to be addressed differently based on the public or private nature of the system.

Table 1.3 applies the attribute list to two operational biometric systems, US-VISIT and Privium [9], to illustrate the classification methodology.

Table 1.3 Application Classifications

US-VISIT	Privium System
Overt	Overt
Noncooperative	Cooperative
Nonhabituated	Nonhabituated
Supervised	Nonsupervised
Nonstandardized	Standardized
Open	Closed
Public	Private

1.8 Summary

Recognition methods that enhance the security of the system and convenience of users have acquired increased importance in today's digital world. Traditional recognition methods based on memorizing secrets or possession of tokens, although still used predominantly, are facing serious operational challenges. Biometric technologies provide an additional level of security and convenience, but this should not be interpreted as biometrics being the perfect solution. Biometric technologies also have limitations. Human interaction plays a significant role in determining the performance of biometric systems, and it has only lately started receiving the attention it deserves. Social acceptance based on geocultural conditions will challenge the user confidence in the technology. Ensuring user privacy is a key factor in increasing the adoption of biometric systems. Biometric systems are not immune to mismatch errors, which are influenced by a variety of factors, including deployment environment, user interaction, and the strength of the underlying biometric matching algorithm. A perfectly secure system has never existed and never will. All systems have vulnerabilities, and a well-designed system should use appropriate combination of knowledge-based, token-based, and biometric technologies to reduce these vulnerabilities. Biometric technologies will play an increasingly larger role in our daily lives, and the rest of this book discusses its various technical aspects, potential applications, challenges, and solutions.

References

[1] NSTC, *Biometrics Glossary*, Washington, D.C., 2006.
[2] Galton, F., *Finger Prints*, London, U.K.: MacMillan, 1892.
[3] Jain, A., and A. Ross, "Introduction to Biometrics," in *Handbook of Biometrics*, A. Jain, P. Flynn, and A. Ross, (eds.), New York: Springer, 2008, pp. 1–22.
[4] ISO, *ISO/IEC 19795-1: Information Technology—Biometric Performance Testing and Reporting—Part 1: Principles and Framework*, Geneva, Switzerland, 2006.
[5] M1.4, *Study Report on Biometrics in E-Authentication*, 2007
[6] Cole, S. A., *Suspect Identities: A History of Fingerprinting and Criminal Identification*, Cambridge, MA: Harvard University Press, 2001.
[7] IBG, *Biometric Market and Industry Report 2009–2014*, New York, 2008.

[8] Wayman, J. L., "Fundamentals of Biometric Authentication Technologies," in *National Biometric Test Center Collected Works*, J. L. Wayman, (ed.), San Jose, CA: San Jose State University, 2000, pp. 11–14.

[9] Schipol, "Schipol—Why Privium?"http://www.schiphol.nl/Travellers/AtSchiphol/Privium-Irisscan/WhyPrivium.html.

CHAPTER 2
Fundamentals of Technical Evaluations

No biometric system is perfect—there are varying levels of how well a biometric system performs its basic task of recognizing users. Evaluating biometric systems requires analyzing a number of different variables such as mismatch error rates, throughput rates, reliability, consistency, cost, and target population. The task of determining which system is the best is made difficult for practitioners and system integrators as they face a deluge of performance reports and press releases about superiority of different systems. In order to infer results of these reports in a correct and unambiguous manner, a clear understanding of biometric system performance evaluation is necessary. This chapter introduces the reader to concepts of performance evaluation and system errors so that questions such as the following can be answered:

- Out of the multiple biometric systems, which one performs the best?
- How well does a particular biometric system perform?
- Why is a particular biometric system performing poorly?

By the end of this chapter, the reader will become familiar with different types of biometric system errors, testing methodologies, performance assessment and visualization techniques, and best practices for planning an evaluation.

2.1 System Process Transactions

Before discussing various types of biometric system errors, it is important to become familiar with concepts and terms that form the basis of performance calculations. As discussed in Chapter 1, a biometric system is capable of three processes: enrollment, verification, and identification. Each of these requires a user to interact with a sensor and provide a sample for further processing, which is called a *presentation* [1]. An operation that consists of a single presentation or a series of presentations that results in an enrollment or a matching score is called an *attempt*. *Transaction* is completion of one or more attempts for the purpose of enrollment, verification, or identification [1]. It is extremely important to understand the concepts of *presentations*, *attempts*, and *transactions* for correct calculation of error rates, as well as differentiating system errors from process errors. To understand the differences between these terms, consider a scenario where a fingerprint recognition system requires a user to provide three usable fingerprint images for enrollment and two usable fingerprint images for verification. Each fingerprint image submitted by the user during enrollment and verification is a presentation. A user provides three fingerprint images in three successive *presentations*, which are all accepted by the system, and thereby completes the enrollment process. This is a single *attempt* and a single *transaction*. During verification the user provides two fingerprint samples, which results in two similarity scores. The aggregate of these two scores is used to make a decision about the identity of the user. In this case two *presentations*, two *attempts,* and one *transaction* have taken place.

Two other important concepts are those of *genuine attempts* and *imposter attempts*. In a genuine attempt a user tries to match his or her sample against his or her own enrollment template. In an imposter attempt a user tries to match his or her sample against another user's enrollment template. The decision policies and the nature of the attempts determine *genuine transactions* and *imposter transactions*. The similarity scores generated from genuine transactions are called *genuine match scores* and similarity scores generated from imposter transactions are called *imposter match scores*. These terms and concepts are used throughout this chapter.

2.2 Types of Errors

A generic biometric system is composed of five subsystems: data acquisition, signal processing, data storage, matching, and decision. Performance assessment of a biometric system is a function of errors generated by these five subsystems, which are discussed in this section.

2.2.1 Acquisition Errors

An acquisition error occurs when the data acquisition subsystem is unable to capture a representation of a user's biometric characteristics or if the signal processing subsystem is unable to extract features from the sample. An acquisition error depends on multiple factors such as user training, user interface and capture device form factor, environment conditions, and sample quality threshold. A user who has not been given adequate training or is unsure about proper interaction with the device may provide an incomplete sample or exceed the system defined time-out threshold. Interaction issues gain even more importance in unsupervised systems as there is no operator to provide corrective feedback to the user. A classic example of an interaction issue is a user not knowing how long to keep his or her finger placed on the sensor. A common issue for first-time users is removing their fingers before the capture process has completed, which leads to an acquisition error. Environmental conditions also have a significant impact on biometric systems, which is discussed in later chapters. Sample quality assessment ensures that low-quality data is kept out of the system and the best possible sample is used for the different processes. Similar to the decision subsystem, quality assessment has a threshold that determines the level of noise allowed in the sample. Biometric samples that do not pass this quality threshold are rejected and users are asked to present another sample.

2.2.2 Matching Errors

The matching subsystem compares two biometric samples and produces a similarity score. The similarity score is compared to a matching threshold and a decision is made about the source of the two samples. Based on this operation, two different types of errors can be committed:

- The inability to correctly assess whether two templates are from the same user;
- The inability to correctly assess whether two templates are from different users.

Matching errors can be attributed to several sources, some of which overlap with acquisition errors. Human factors, improper user training, and environmental factors impact the consistency of samples captured by the sensor, which in turn impacts the matching process. The matching of two samples is typically performed on samples that are captured with a time gap between them; for example, a user creates a fingerprint enrollment template today and then returns after a week and provides the same fingerprint for matching purpose. In the course of a week the user may have scarred his or her finger, which temporarily alters the fingerprint representation. This can lead to a mismatch error.

2.3 Performance Metrics

In this section the fundamental system errors discussed in Section 2.2 are described in terms of error rates and the proportion of users affected by the errors.

2.3.1 Failure to Enroll Rate (FTE)

The FTE is the proportion of user enrollment transactions that cannot be completed according to the enrollment policy [1]. The root cause of an FTE could be any of the ones described in Section 2.2.1 and is governed by administrative policy decisions. For example, consider an enrollment policy that allows the user three successive attempts in order to complete the process by presenting an acceptable biometric sample. If a user cannot present an acceptable biometric sample in three attempts, then it is considered a failure to complete the enrollment process. The number of attempts in this case is purely a policy decision.

2.3.2 Failure to Acquire Rate (FTA)

The FTA is the probability of user attempts during identification or verification for which the system cannot acquire an appropriate sample [1]. Even though the root cause for FTA and FTE could be the same, they are differentiated based on the process during which the error occurs. This ensures that errors are not counted twice and are attributed to the appropriate process as opposed to a particular subsystem.

2.3.3 False Nonmatch Rate (FNMR)

The false nonmatch rate (FNMR) is calculated as the proportion of samples from genuine attempts that cannot be matched against the enrolled templates of genuine users [2].

$$FNMR = \frac{\text{Number of rejected genuine comparisons}}{\text{Total number of genuine comparisons}} \tag{2.1}$$

2.3.4 False Match Rate (FMR)

The FMR is calculated as the proportion of samples from imposter attempts that are successfully matched against the enrolled templates of genuine users [2].

$$FMR = \frac{\text{Number of imposter comparisons}}{\text{Total number of imposter comparisons}} \tag{2.2}$$

The root cause of both false match and false nonmatch errors can be traced to matching errors described in Section 2.2.2. It should be noted these error rates are calculated based on a single attempt and not on a verification or identification transaction. Verification and identification transaction errors are described in Sections 2.3.5 and 2.3.6.

2.3.5 Verification Performance Metrics

During verification a user makes a claim to an identity and the input sample is compared to the associated enrolled template for the claimed identity. The output of the verification transaction is either an acceptance or a rejection of the claim, and the errors based on these claims are described next.

2.3.5.1 False Reject Rate (FRR)

The FRR is calculated as the proportion of verification transactions from genuine users that will be incorrectly rejected [2]. For single-attempt transactions, the FRR includes the FTA. Thus, the formula for calculating FRR is given by

$$FRR = FTA + FNMR * (1 - FTA) \tag{2.3}$$

2.3.5.2 False Accept Rate (FAR)

The FAR is calculated as the proportion of verification transactions from imposters that will be incorrectly accepted [2]. For single-attempt transactions a false accept occurs only if a false match occurs without a failure to acquire error. Thus, the formula for calculating FAR is given by

$$FAR = FMR * (1 - FTA) \tag{2.4}$$

Both FAR and FRR calculations are based on transactions. As discussed earlier, a transaction is a policy decision that takes into account the number of failed attempts. Consider a biometric system that allows a user three failed verification attempts before rejecting the user. In the case of a failed verification transaction, there are three false nonmatch errors, but only one false reject error will have taken place. This distinction is extremely important for the calculation of system errors such as FMR and FNMR and process errors such as FAR and FRR.

2.3.5.3 Equal Error Rate (EER)

The EER is calculated as the point where the FAR and FRR are equal. This rate is also called the *crossover error rate*. A lower EER indicates a better overall matching performance. The EER is often used for comparing multiple biometric systems, but it does not provide much benefit for operational evaluation because both error rates are given equal weighting.

2.3.5.4 Generalized Error Rates

The FAR and FRR take into account the FTE as part of the final error rate. The FAR and FRR of a system can be positively influenced by increasing the FTE of a system and keeping problematic users out of the system. In cases where multiple systems need to be compared, a direct comparison of the FAR and FRR will not highlight the true differences in system capability. The generalization of error rates is then necessary for comparing multiple systems with a different FTE. Generalized error rates for the FAR and FRR, *GFAR* and *GFRR*, are calculated by combining enrollment, acquisition, and matching error rates. Every instance of an FTE is treated as a successful enrollment, but every subsequent verification attempt against or by the user is treated as an error.

The GFAR is described as a proportion of verification transactions from imposters that will be successfully enrolled and incorrectly accepted by the system.

$$GFAR = (1 - FTE) * FAR \tag{2.5}$$

The GFRR is described as proportion of verification transactions from genuine users who will be incorrectly rejected, will fail to enroll, or will not be successfully acquired for verification.

$$GFRR = FTE + (1 - FTE) * FRR \tag{2.6}$$

2.3.6 Identification Performance Metrics

As discussed in Chapter 1, the output of identification can be a candidate list of enrolled users who are the most similar to the input sample. The identification rank r of a user is the smallest-sized candidate list of which the user is a member. As a part of best practices, the total number of users enrolled in a database is also mentioned along with the rank of a user. For example, if the total number of enrolled users in a database is n, the user's rank is presented as rank r out of n.

The closed-set identification performance is described with respect to rank, as the input sample belongs to an enrolled user. The identification rate at rank r is the probability that the enrolled user is a member of the candidate list of size r for an identification transaction. *Cumulative match characteristic* (CMC) curves are generally used to report the closed-set identification performance, which is described in Section 2.5.

In open-set identification the input sample could potentially belong to a nonenrolled user. *The false negative identification rate* (FNIR) and *false positive identification rate* (FPIR) are metrics used to describe open-set identification performance. The *FNIR* is the proportion of identification transactions performed by enrolled users that returns a candidate list of which they are not a member. The *FPIR* is the proportion of identification transactions performed by nonenrolled users that returns a candidate list of which they are a member. For example, the FPIR is the probability of an innocent traveler being on a candidate list when his or her biometric is compared against a watchlist.

$$FNIR = FTA + (1 - FTA) * FNMR \qquad (2.7)$$

$$FPIR = (1 - FTA) * \left[1 - (1 - FMR)^n\right] \qquad (2.8)$$

where n is the number of enrolled templates.

2.4 Type I and Type II Errors

Statistical hypothesis testing forms the basis of evaluating biometric system matching errors. In statistics hypothesis testing is conducted to test an assumption or a claim. For every assumption, a null hypothesis and a corresponding alternate hypothesis are generated. For a biometric system a null hypothesis states that two samples being compared belong to the same individual, and the alternate hypothesis states that the two samples being compared belong to different individuals:

Null hypothesis H_0: samples belong to same individual P
Alternate hypothesis H_1: samples do not belong to the same individual P

If two samples that are being compared belong to the same individual but are determined to come from different individuals, then the null hypothesis is rejected and an error is committed. Such an error is called a Type I error. If the two samples that are being compared belong to different individuals but are determined to belong to the same individual, then the alternate hypothesis is accepted and an error is committed. Such an error is a Type II error. A false reject or false nonmatch error is analogous to the Type I error, and a false accept or false match error is analogous to the Type II error. The genuine and imposter comparisons produce match scores that can be represented by genuine and imposter score distributions. The ultimate goal in biometric system design is to create a system that produces genuine and imposter distributions that do not overlap because the area of overlap signifies the total proportion of errors produced by the system. The threshold decides the proportion of errors that are categorized into false accepts and false rejects. Figure 2.1 illustrates this concept. By moving the threshold to the right, the proportion of false accepts decreases and the proportion of false rejects increases. This is the underlying principle that governs the trade-off between security and the convenience of a biometric system, which is discussed in the next section.

2.5 Performance Curves

Previously discussed error rates provide the capability of assessing a single error category such as false accepts or false rejects. A biometric system is designed to maximize *security* and *convenience*. For a biometric system security is the ability of the system to detect imposter transactions reliably and accurately, whereas convenience is the ability of the system to detect genuine transactions reliably and accurately. A more detailed and realistic analysis of performance measures requires it to be viewed as a function of *security* and *convenience*. *Detection error trade-off* (DET) curves and *cumulative match characteristic* (CMC) curves are two such analytical methods used predominantly in the biometrics domain for performance analysis. A DET curve plots the false match and false nonmatch rates on the x-axis and the y-axis as a function of the threshold. Thus, the DET curve graphically presents the trade-off between the two error rates. For each possible threshold value the two error rates are calculated and plotted on the graph. Thus, each point (x, y) represents the combination of error rates at a particular threshold value. A DET curve can also plot the FAR and FRR on the x-axis and the y-axis as a function of the threshold, and this represents a trade-off in the transaction errors. Figure 2.2 illustrates a DET curve. The closer the curve is to the origin, the better is the performance of the system. An ideal biometric system with no errors will have a DET curve that is a straight line on the x-axis or the y-axis. DET curves are often plotted

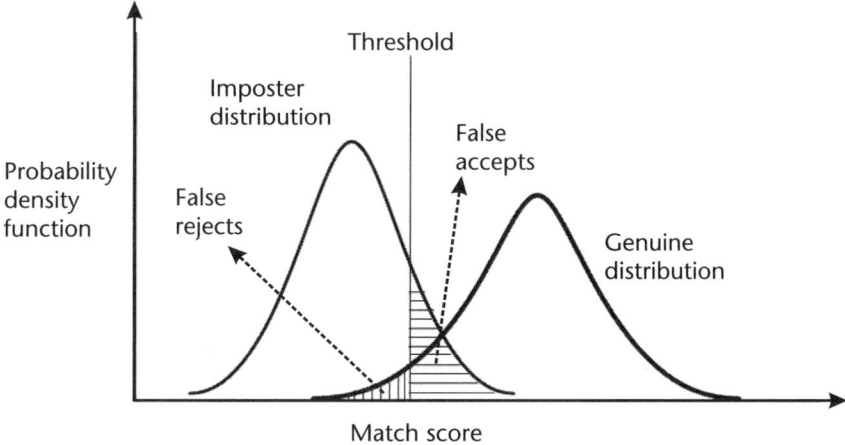

Figure 2.1 Biometric system match score distributions.

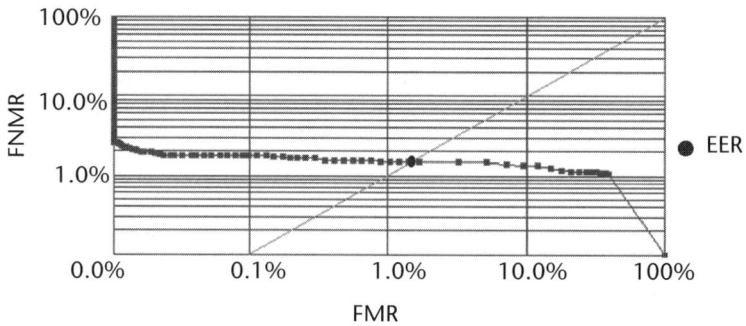

Figure 2.2 DET curve.

on a logarithmic scale on both the axes. This provides visual clarity to the graph lower error rates, which in turn allows the viewer to visualize the trade-off in much greater detail.

Identification results are visually analyzed using a *cumulative match characteristic* (CMC) curve. The x-axis of a CMC curve represents all possible rank values, and the y-axis represents the probability of correct identification at each possible rank value. The CMC curve in Figure 2.3 graphically illustrates the identification accuracy of the biometric system against a variable sized candidate list.

The analytical power of DET and CMC curves can be used to either analyze a single system or compare multiple systems. For instance, an administrator may already have decided which biometric system to use but is undecided about the optimal threshold for the biometric system. In such a case the administrator can analyze a DET curve and determine the optimal threshold based on acceptable trade-off between the two error rates. In another scenario an administrator might want to compare multiple biometric systems and decide which one to select for deployment. In such a case multiple DET curves can be superimposed on the same graph and compared using a common scale. In Figure 2.4 DET curves for systems S1, S2, and S3 are shown superimposed on the same graph. Figure 2.5 shows CMC curves for four systems superimposed for comparion purposes. A single CMC curve or multiple CMC curves can be analyzed using the same methodology.

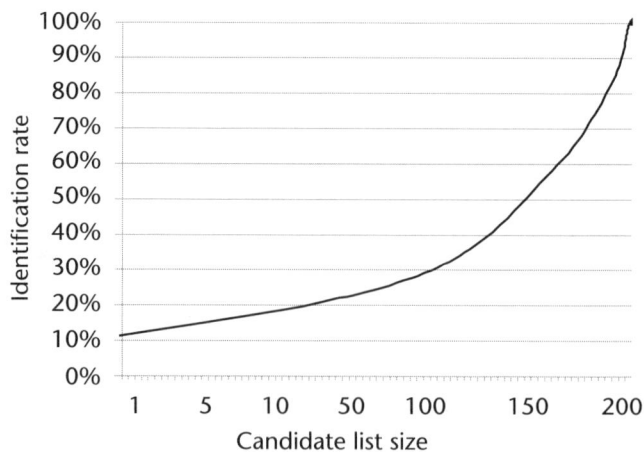

Figure 2.3 CMC curve.

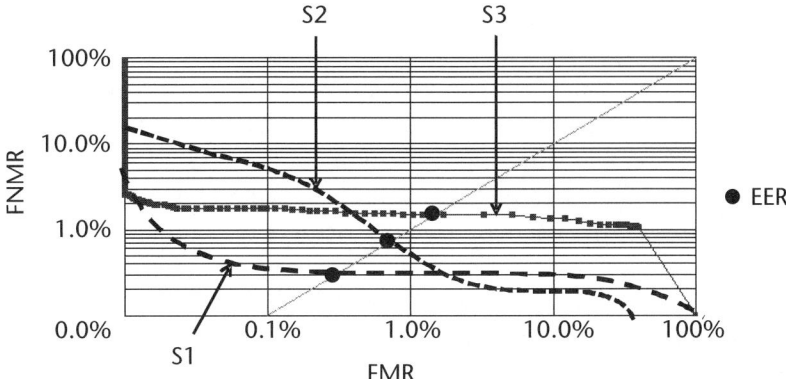

Figure 2.4 Comparison of multiple DET curves.

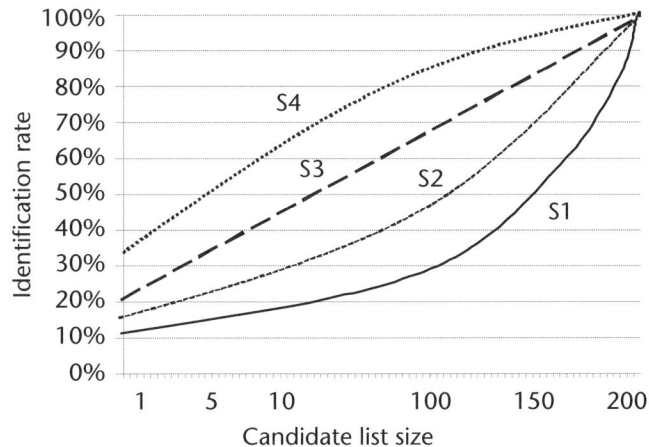

Figure 2.5 Comparison of multiple CMC curves.

2.6 User-Specific Performance: Zoo Analysis

Doddington et al., as part of their research on speaker verification, devised an alternative method of analyzing biometric system performance [3]. Instead of concentrating on fundamental system error rates and presenting the performance of a system in terms of overall error rates, their approach analyzed genuine match scores and imposter match scores of users in a given system. In Doddington's zoo model all users of a biometric system are categorized into one of four animal groups based on the similarity scores they generate.

1. *Lambs:* Users belonging to this group are easy to imitate. Imposters matching against the templates of lambs generate a high similarity score. Lambs have an adverse impact on biometric systems, as they generate a high proportion of false matches.
2. *Wolves:* Users belonging to this group easily imitate users belonging to other groups. Wolves' templates generate a high similarity score when matched

against other users' templates. They have an adverse impact on biometric systems, as they generate a high proportion of false matches.
3. *Goats:* Users belonging to this group are difficult to match against themselves. They have an adverse impact on biometric systems, as they generate a high proportion of false nonmatches.
4. *Sheep:* Users belonging to this group match well against themselves. A majority of the users of a biometric system belong to this group.

A zoo plot can be visualized by calculating the average genuine matching scores and average imposter matching scores for all users of a biometric system and plotting them on 2-D graph of genuine match scores versus imposter match scores. The resulting animal groups separate into four different quadrants of the zoo plot, as shown in Figure 2.6.

Doddington's zoo analysis uses genuine match scores or imposter match scores to identify issues and suggest improvements. Dunstone and Yager added four more animals to the original zoo, which uses the relationship between genuine match scores and imposter match scores for categorizing users [4]. These four animal groups are:

1. *Chameleons:* Users belonging to this group always appear similar to themselves and other users. They generate a high similarity score, irrespective of whom they are matched against, thereby generating false matches.
2. *Phantoms:* Users belonging to this group do not appear similar to themselves or to other users. They generate a low similarity score, irrespective of whom they are matched against, thereby generating false nonmatches.
3. *Worms:* Users belonging to this group appear similar to other users but do not appear similar to themselves. They generate a low similarity score when matched against their own templates and generate a high similarity score when matched against users of other groups. They generate false matches and false nonmatches and represent the most problematic user group of the biometric system.

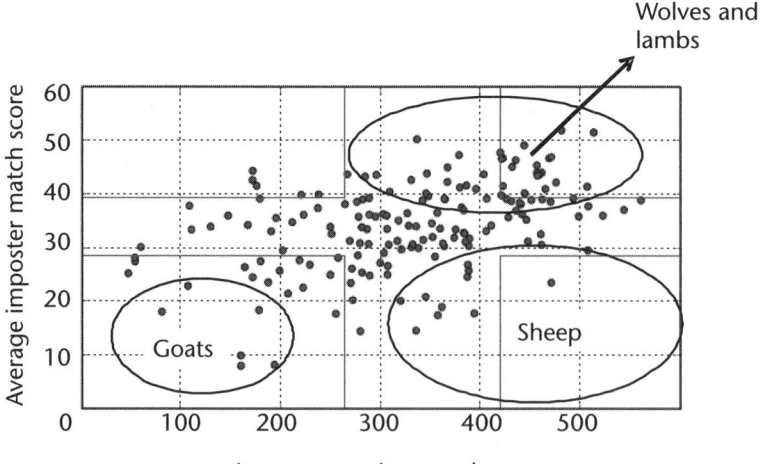

Figure 2.6 Doddington's zoo analysis.

Free Subscription
Artech Direct email newsletter

New Title News • Special Offers • Author Insights

☐ Yes! Please enter my free subscription to *Artech Direct* and keep me up-to-date with emailed news of product and service information from Artech House/Horizon House Publishers.

email address: _____

You may also make my email address available to selected industry organizations and companies. ☐ Yes ☐ No

Please indicate your areas of interest	
☐ Telecommunications/Wireless/Networking	☐ Building Technology
☐ Software Engineering/Computer Security	☐ Signal Processing
☐ Microwave	☐ Sensors/MEMS/Nanotechnology
☐ Radar/Remote Sensing/Electronic Defense	☐ Antennas & Propagation
☐ Power Engineering	☐ Engineering Management
	☐ Biomedical Engineering

Mailing address:

Name: _____

Company: _____

Address: _____

_____

Fax or mail this card to the Artech House office nearest you. Please see other side.

ARTECH HOUSE BOSTON | LONDON

To see the full line of Artech House books and software, visit our online bookstore at:

www.artechhouse.com

Special Offers • Sample Chapters • Secure Ordering

To receive information on new and forthcoming titles from Artech House, please fill out the other side of this card and mail or fax it to one of the locations below:

For Europe, Asia, Middle East, Africa:

Artech House
16 Sussex Street
London, SW1V 4RW U.K.
+44 (0)20 7596 8750
FAX: +44 (0)20 7630 0166
artech-uk@artechhouse.com

All other regions:

Artech House
685 Canton Street
Norwood, MA 02062 U.S.A.
1-781-769-9750
1-800-225-9977 (continental U.S. only)
FAX: 1-781-769-6334
artech@artechhouse.com

ARTECH HOUSE BOSTON | LONDON

4. *Doves:* Users belonging to this group appear similar to themselves, but do not appear similar to other users. They generate high similarity scores when matched against their own templates and generate low similarity scores when matched against users of other groups. Users of this group represent the most ideal users of the biometric system.

The zoo analysis in its entirety provides a framework for isolating user-specific issues, which is not possible using global error rate analysis. For example, if a specific user or group of users is contributing disproportionately to the error rates, zoo analysis can identify such a group of users and preventive measures can be taken that only affect that group of users and not all users of a system. This is the typical problem scenario that most biometric system administrators have to address in an operational system.

2.7 Evaluation Methodologies

Just as software and hardware engineering have specific testing methodologies based on the specific objectives, biometric system performance is evaluated using three different testing methodologies: technology evaluation, scenario evaluation, and operational evaluation [5]. Table 2.1 summarizes the differences in these three methods.

One very important factor that sets the evaluation of biometric systems apart from the evaluation of other security systems, such as cryptographic systems, is the need to use biometric data collected from humans. Although biometric data modeling has made great strides in the last decade, a true evaluation of a biometric system still requires data collected from live humans. This increases the complexity and the financial cost of conducting biometric system evaluations.

Table 2.1 Comparison of Technology, Scenario, and Operational Evaluation

Characteristic	*Technology*	*Scenario*	*Operational*
Scope of test	Individual subsystems, combination of subsystems	Complete system	Complete system
Data processing	Off-line	Online or off-line	Online
Repeatable	Possible	Depends on level of control	Not possible
Comparable	Possible	Depends on level of control	Not possible
Ground truth	Known	Known	Generally unknown
Real-time feedback	Not possible	Possible	Possible
Test supervisor	Required	Application dependent	Application dependent
Level of control (environment, users)	High	Medium	Low
Performance metrics	FTA, FTE, FMR, FNMR, FAR, FRR	FTA, FTE, FMR, FNMR, FAR, FRR, throughput	Throughput; other metrics require ground truth

Source: [1].

2.7.1 Technology Evaluation

The goal of technology evaluation is to evaluate any of the biometric subsystems for a target application. Typically a technology evaluation is conducted to compare multiple biometric subsystems from different vendors. A well-known example of a technology evaluation is the Fingerprint Verification Competition (FVC) [6]. The goal of the FVC is to test multiple fingerprint recognition algorithms on the same set of fingerprints collected from specific fingerprint sensors. Such an evaluation provides several benefits. Data collection and data processing are conducted separately, thus providing a complete dataset for cross comparisons. The ground truth of the all data is known so that the true identity is linked with all biometric samples. Since there is no requirement for such a system to operate in real time, it is well suited to laboratory research. Technology evaluation can also be used to evaluate changes in multiple versions of the same algorithm as well.

2.7.2 Scenario Evaluation

The goal of a scenario evaluation is to determine the overall performance of a biometric system in a simulated application that is representative of the real-world application. A scenario evaluation includes the entire biometric system from the data acquisition subsystem to the decision subsystem. Scenario evaluations are conducted in real time, although certain tests separate data collection from data processing. The key benefit of a scenario evaluation is the inclusion of human interaction variability and the ability to conduct enrollment, verification, and identification transactions. As discussed earlier, transactions are a product of policy decisions, and a scenario evaluation provides a method for evaluating different policy decisions. If all the transactions are completed in real time, throughput analysis is also possible.

A scenario evaluation does not allow a true comparison of multiple systems as the biometric data used for each evaluation is not collected from the same sensor. Environment conditions, human interaction issues, and subject demographic differences have an impact on the eventual performance of the system, and it is impossible to replicate all of these among multiple systems. A scenario evaluation is useful for understanding impact of real-world factors on the system performance without having to deploy the system. Like technology evaluations, the ground truth of data is known, which allows a deeper analysis on the matching errors.

2.7.3 Operational Evaluation

The goal of an operational evaluation is to determine performance of a complete biometric system deployed in a real-world application that is being used by a particular user population. Such an evaluation can be viewed as a monitoring and maintenance activity. Results of operational evaluations are nonrepeatable, so a comparison of multiple operational evaluations is impractical. For example, a comparison of operational evaluations of the same face recognition system at an airport and a seaport will provide different results as the environmental effects and the demographics of the users for the two applications are completely different. In operational evaluations the ground truth cannot be determined, which limits the amount of analysis that can be conducted on matching errors. For example, the system being evaluated could have been functioning for a long time, and in such a case the enrollments have already been performed before the evaluation commenced. Also, evaluators do not have any control over the target population, which gives them less control over specific manipulation of test variables. Operational evaluations can be extremely useful for conducting user experience and throughput analysis. Users are more likely to provide an honest opinion of their experiences in an operational evaluation compared to a laboratory or a simulated test. Throughput analysis is also more realistically determined when the biometric system is integrated into the operational infrastructure. For example, the throughput of face recognition integrated with a turnstile system will be different than the throughput of face recognition used as part of the network logon.

2.8 Design of Evaluation

Previous sections of this chapter discussed performance metrics and evaluation methodologies as they relate to biometric systems. This information is necessary for designing effective evaluations and reporting results of these evaluations. This is an important section even for readers who might never conduct an evaluation, as it discusses the process behind an evaluation and provides readers with the skills to ask pertinent questions when analyzing other system evaluation reports.

One should think of a biometric system evaluation as a scientific experiment where a system is treated with several carefully chosen input and the effect of this input on the output is observed. As with any other experiment, a couple of key questions need to be answered before starting an evaluation:

- What is/are the main factor/factors being evaluated?
- Which evaluation methodology (technology, scenario, operational) is the most suitable to achieve the overall objective of the evaluation?

The answers to these questions will determine specific details such as the number of genuine transactions and imposter transactions, number of subjects, data collection procedures, calculation of performance metrics, and information to be collected from the system.

A well-designed evaluation requires determining which variables have an impact on performance and how to control them. These factors are generally categorized into five groups [7]. Readers familiar with the design of experiments will notice a direct correspondence with the following groups:

1. Factors that are observed;
2. Factors that are manipulated to see what effect they have on the observed factors;
3. Factors that are controlled so that they have a negligible effect on the evaluation;
4. Factors that are randomized to minimize their impact on the evaluation;
5. Factors that have a negligible effect and are ignored.

The determination of how to categorize various factors into these groups requires expertise in data collection, design of experiments, statistical evaluation, and, most importantly, the particular biometric technology under observation. The ISO/IEC 19795-1 [1] lists several factors that have an impact on performance, although it should be noted that this list is not exhaustive, nor does each factor affect all biometric technologies. Table 2.2 lists the general factors that have an impact on biometric system performance.

2.8.1 Sample Quality

Sample quality assessment is an extremely important step in biometric system evaluations. Sample quality is impacted by the inherent characteristics of the user, the system deployment characteristics, the deployment environment, the interaction between the user and system, or a combination of any of them. Quality is an extremely subjective concept and it depends on the context in which it is being used. Within the biometrics domain, the term quality is used in three different contexts [8]:

1. *Fidelity* reflects the accuracy of a sample's representation of the original source.
2. *Character* reflects the expression of the inherent features of the source.
3. *Utility* reflects the observed or predicted positive or negative contribution of the biometric sample to the overall performance of a biometric system.

The ISO/IEC 29794-1:2009 standard [8] specifies a modality independent quality derivation and interpretation framework. The standard prescribes that quality score should be indicative of performance metrics such as the FTA, FTE, FMR, and FNMR. Utility, as described earlier, is a function of both fidelity and character and predictive of system performance. Extensive research is being conducted in this area to identify covariates that can be used in calculating the utility of a biometric sample.

A discussion of different variables that have an impact on biometric technologies is given in each specific technology chapter (Chapters 3 through 10), along with a discussion of its impact on recognition performance.

Table 2.2 Factors Affecting Biometric System Performance

Factors	Examples/Description
Population demographics	Age, gender, occupation, ethnic origin
Application	Time between transactions, identification or verification processes, number of attempts per transaction, supervised or unsupervised system
User physiology	Physical appearance such as height, baldness, height, color of iris, skin tone, and cold and sore throat
User interface	Physical or nonphysical interaction, auditory or visual feedback, number of interactions for a complete transaction
Environmental influences	Temperature, ambient lighting, humidity, rain, and snow
Data acquisition sensors	Dirt, residue, obstruction, form factor
Database size	Number of individuals in the enrolled database

Source: [2].

2.9 Reporting Biometric Evaluations

Performance evaluations are heavily dependent on the type of biometric system, the application context, the deployment environment, the target population, the number of users participating in the test, and other factors. A properly formulated report provides the reader with accurate information regarding the evaluation so that it can be interpreted correctly without any ambiguity. Table 2.3 lists the areas that are critical as part of a comprehensive performance evaluation report.

2.10 Summary

This chapter introduces readers to various topics related to the performance evaluation of biometric systems including fundamental system errors and error rates, transactional error rates, graphical techniques for analyzing these error rates, and system evaluation methodologies. Performance evaluations are critical for successful deployments; it gives decision-makers information at their disposal to make educated decisions about procurement, system administrators can fine-tune performance based on specific application context and predict future performance, and vendors can identify performance issues that need to be addressed. The goal of this chapter is to lay a solid foundation for conducting performance evaluations of specific biometric modalities. Biometric technologies have advanced significantly in the last decade and their use in specific applications will increase in the near future. The ability to conduct meaningful comparisons and assessments will be crucial to successful deployments and increasing biometric adoption.

Table 2.3 Components of an Evaluation Report

Area	Description
Details of system tested	Entire biometric system or a particular component of the entire system
Type of evaluation	Technology, scenario, or operation
Design details	Explanation of data collection protocol, and categorization of various factors as described in Section 2.8
Details of test environment	Physical layout of collection area, environment conditions, time of year
Crew size	Number of individuals who participated in the data collection along with the breakdown of genuine and imposter groups
Demographics of crew	Breakdown of age, gender, occupation, and other relevant information
Thresholds	Sample quality and matching threshold
Transaction policy	Number of transactions for enrollment, verification, or identification along with time difference between presentations and attempts
Performance metrics	Justification for calculation of specific performance metrics and formula for calculating them
Deviations	Any abnormalities and outliers should be reported; for example, if any particular user was discarded from the evaluation, a reason for doing so should be present in the report

Source: [2].

References

[1] ISO, *ISO/IEC 19795-1: Information Technology—Biometric Performance Testing and Reporting—Part 1: Principles and Framework*, Geneva, Switzerland, 2006.

[2] Wayman, J. L., and T. J. Mansfield, *Best Practices in Testing and Reporting Performance of Biometric Devices*, Middlesex, U.K.: National Physical Laboratory, 2002.

[3] Doddington, G., et al., "The DET Curve in Assessment of Detection Task Performance," *Fifth European Conference on Speed Communication and Technology*, Greece, 1997, pp. 1895–1898.

[4] Dunstone, T., and N. Yager, *Design, Evaluation, and Data Mining*, New York: Springer-Verlag, 2008.

[5] Phillips, P. J., et al., "An Introduction to Evaluating Biometric Systems," *Computer*, 2000, pp. 56–63.

[6] Maio, D., et al., "FVC2000: Fingerprint Verification Competition," *IEEE Transactions on Pattern Machine Intelligence*, Vol. 24, March 2002, pp. 402–412.

[7] Montgomery, D. C., *Design and Analysis of Experiments*, New York: John Wiley & Sons, 2000.

[8] ISO/IEC, *ISO/IEC 29794-1:2009 Information Technology—Biometric Sample Quality—Part 1: Framework*, Geneva, Switzerland, 2009.

CHAPTER 3
Fingerprint Recognition

Fingerprint recognition, also known as dermatoglyphics, is the oldest and most widely deployed biometric technology. A close observation of the skin surface on fingers reveals a flow of raised lines called friction ridges. A fingerprint is the resulting impression made by these ridges. Over a century of research and real-world use has shown that the information contained in fingerprints is unique to each individual and is an extremely effective method of recognition. From its roots in law enforcement, fingerprint recognition is now deployed in practically every type of government and commercial application that uses biometrics. Fingerprint recognition has a vast body of knowledge compared to other biometric technologies and it would be impossible to compress all of it into this chapter. This chapter will discuss the underlying basics of the technology, various components of the technology, its practical applications, and design and deployment considerations.

3.1 Fingerprint Anatomy

If you look closely at the skin of your fingers and palm, you will notice that the skin forms a pattern of ridges and valleys, as seen in Figure 3.1. These ridges are called friction or papillary ridges. The flow of ridges is not continuous; the ridge lines abruptly end or diverge. The point where there is a discontinuity in the ridge line is called a *minutiae point* and a majority of fingerprint recognition algorithms use minutiae points for matching purposes. These ridges are fully formed by the seventh month of fetus development and remain unchanged for the life of the individual [1]. This pattern is formed at random during fetus development, and thus, even twins have different fingerprints [2]. The skin is made up of two main layers: the *epidermis* (outer layer) and *dermis* (inner layer). The ridge lines constitute the epidermis, whereas sweat glands, blood vessels, nerves, and other cellular structures are present in the dermis. The ridge structure is a representation of the arrangement of various cell structures in the dermis. Thus, even if the epidermis is damaged or scarred, the ridges will recover and retain the original pattern when the epidermis heals. This dual property of permanence and recovery makes fingerprints a good candidate for recognition.

Figure 3.1 Fingerprint.

3.2 History

Archaeological evidence dating back to 6000 B.C. has shown the use of fingerprints on clay pottery to mark the identity of the potter [3]. Bricks used in the ancient city of Jericho have been discovered with impression of fingerprints, and its most probable use was to recognize the mason. The earliest scientific studies on fingerprints were conducted by two microscopists named Govard Bidloo and Macello Malphighi, who published their findings and conclusions on friction ridges in 1685 and 1687, respectively [4]. The main point of discussion in those publications was whether friction ridges were organs of touch or if they facilitated sweating. Then in 1778 anatomist J. C. Mayer made the first recorded claim about the uniqueness of fingerprints by suggesting in his publication that the arrangement of skin ridges could never be duplicated in two persons. In 1823 Czech physician Jan Purkyne classified the structure of friction ridges into nine categories in an effort to establish a connection between vision and touch [5]. Although Purkyne was a trained physician, he was also a student of philosophy and believed that every natural object is identical to itself and claimed that no two fingerprint patterns were identical. In 1858 William Herschel asked a road contractor to impress his handprint in ink on a deed with the intention of adding nonrepudiation to the contract. Although this was a promising technique, it was impractical to catalog all the handprints and check each one manually. In 1880 Henry Faulds applied the idea of fingerprint recognition to identify criminals, which was the first foray of this method for criminal identification. Faulds devised an alphabet classification scheme so that a person's set of 10 prints could be represented by a word that could then be cataloged in a dictionary for quick searching. Faulds published his observations and methodology in a letter to *Nature* in 1880 and also approached Scotland Yard with his technique, but it was perceived to be impractical and subsequently rejected. In 1880 Francis Galton followed up on Faulds' experiment and pursued the previous path of cataloging fingerprints based on the global flow of the ridgelines and ran into the same issues of impracticality. In the course of his experiments he noticed several features of ridges, shown in Figure 3.2, including friction ridge ending and bifurcation, which are predominantly used in automated matchers. Galton then proposed that two fingerprints could be matched by comparing these discontinuities, which

Ending   Bifurcation   Lake/enclosure   Hook   Crossover   Dot

Figure 3.2 Galton minutiae features [6].

he called "minutiae" [6]. Although theoretically this method had merit, Galton believed that it was impractical since the process of matching minutiae would be extremely time-consuming. This discovery would not be used to its fullest until the advent of automated minutiae matchers in the 1960s.

3.2.1 Henry Classification

Edward Henry was a colonial police officer in India interested in using fingerprints to identify criminals since he believed the Bertillon system based on anthropometrics was inadequate [7]. Henry and his assistants set about creating a feasible and practical fingerprint classification scheme and devised a solution that used the global ridge flow of the fingerprint and combined it with the finger number. Their classification methodology had 1,204 primary groupings and was capable of accommodating up to 100,000 prints [8]. Henry introduced this system in his jurisdiction in 1895, and by 1897 it was adopted all over India. Henry published his system in a publication called *The Classification and Uses of Finger Prints* in 1900 [9], and by 1902 Scotland Yard had fully adopted fingerprinting for purposes of identification. Until the advent of automated minutiae-based matching, the Henry system formed the basis of fingerprint recognition in law enforcement applications. Readers who are interested in the details of the Henry classification are encouraged to read [10]. Figure 3.3 illustrates the five major classes of fingerprints used in the Henry classification. The Henry classification system was a watershed moment in fingerprint recognition and early automated systems attempted to emulate this system.

3.2.2 Automated Fingerprint Recognition

Attempts to automate the process of identification can be found dating back to 1920. IBM punch card sorters and other automated data processing technologies

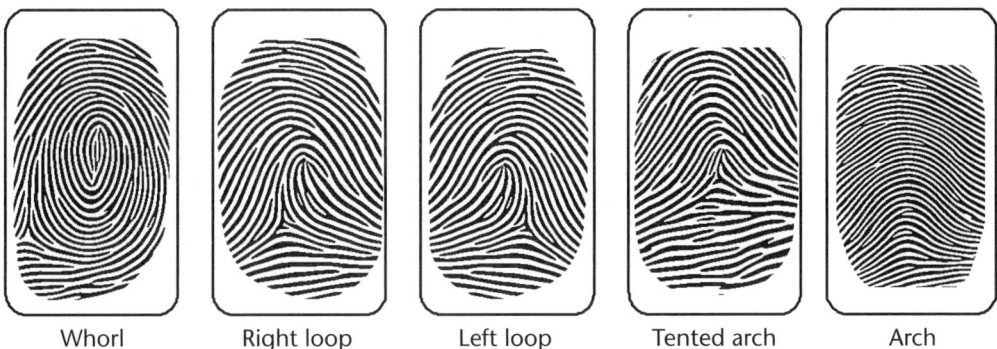

Whorl   Right loop   Left loop   Tented arch   Arch

Figure 3.3 Henry classification.

were employed to alleviate the problem of handling large number of cards. These approaches solved part of the problem, but human examiners were still required to inspect fingerprints. The need to automate the entire matching process was widely acknowledged, but technology constraints remained a barrier to progress. In 1963 Joseph Wegstein and Raymond Moore began work on an automated fingerprint identification system under the auspices of the Federal Bureau of Investigation (FBI) and the National Bureau of Standards (now renamed NIST) [11]. In 1972 the FBI installed a system with a fingerprint scanner built by Cornell Aeronautical Laboratory and a prototype fingerprint reader system built by North American Aviation [4]. With considerable progress achieved on digitizing and automating the matching process, the FBI started scanning all their fingerprint records for persons born after January 1, 1929, and by 1980 they had a databank of 14.3 million records. Throughout the 1980s various city police departments and state criminal justice bureaus in the United States started deploying Automated Fingerprint Identification Systems (AFIS) confirming the continuing importance of the technology.

3.3 Fingerprint Presentation and Acquisition

Fingerprint acquisition methods have evolved from the inking method used from the mid-1800s to the 1960s when the first generation automated fingerprint recognition systems were designed to capture digitized fingerprint images, which are also called *live scan systems*. Fingerprint images captured by live scan technologies are generally described by capture resolution, capture area, image contrast, bit depth level, and geometric distortion [12]. This section will discuss various acquisition methods and technologies.

3.3.1 Inked Capture

This is the oldest form of fingerprint acquisition. Ink is applied on the skin of the finger and an impression of the fingerprint is captured on a physical substrate. A specially trained operator oversees the capture process as the resulting fingerprint impression can be impacted by the amount of ink on the finger and the pressure applied by the finger on the substrate. If the wrong amount of ink is applied or the

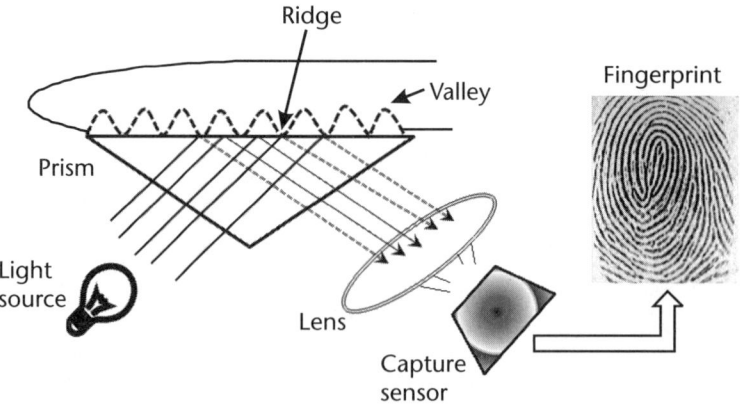

Figure 3.4 Optical fingerprint capture.

wrong amount of pressure is applied, the resulting fingerprint impression might be too light or too dark and pose a challenge for feature extraction and matching. There are several different types of inks available, some of which do not leave any visible trace of ink residue on the finger. Inked capture is predominantly used in law enforcement, where the inked prints are digitized using a scanner, but its use is declining as live scan capture technologies become less expensive.

3.3.2 Latent Fingerprints

Latent fingerprints are formed from the residue left behind on a surface where a finger has come in contact with it. The oily nature of the skin leaves traces of the ridgeline structure, and these traces can be transferred onto a substrate using different types of chemicals. Latent fingerprints are used in forensic science and criminalistics, and readers who have seen a movie or a television show involving a crime scene analysis will have seen the process of capturing a latent fingerprint. This process requires a high degree of specialization to maintain the integrity of the fingerprint.

3.3.3 Optical Sensors

Optical fingerprint capture devices have the longest history of all automated fingerprint image acquisition devices [3]. The earliest methods of optical capture involved using a camera and taking direct images of fingerprints. This method had several issues; the most prominent one was that fingerprint ridges and valleys were not differentiated by color and a shadow needed to be introduced to differentiate ridges and valleys [13]. The next major advancement in this technology was the introduction of the frustrated total internal reflection (FTIR). The refractive index is different for ridges and valleys when a finger touches the platen of the optical sensor. Due to this phenomenon, light incident on valleys is totally reflected and the light incident on ridges is not reflected, which results in the ridges appearing dark in the final image. Older generation optical sensors used CCD cameras to capture the resulting image, but newer generation optical sensors used CMOS cameras. Several commercially available sensors use multiple mirrors and sheet prisms to reduce the size of sensors. By introducing reflective surfaces, the total optical length between the finger surface and the camera is reduced, thereby reducing the size of the sensor.

3.3.4 Solid State Capacitive Sensors

Capacitive sensors work on the principle that skin on finger surface is an equipotential surface; the constancy of potential is a requirement to obtain representative fingerprint images [14]. Capacitive sensors are constructed using a two-dimensional array of conductive plates. The finger is placed on a surface above the array so that the electrical capacitance of these plates is affected, as illustrated in Figure 3.5. The sensor plates under the ridge will have a larger capacitance than the sensor plates beneath the valley. This is because air has a lower permittivity than skin, which leads to an increased capacitance in plates under the skin.

Capacitive sensors can capture fingerprints using two different interaction mechanisms: touch and swipe. In a touch sensor an individual will place his or

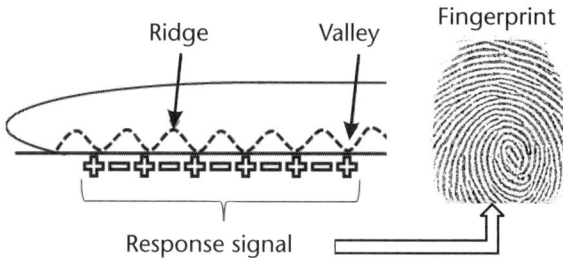

Figure 3.5 Touch capacitive sensor.

her finger on sensor and it will capture the impression in its entirety. In a swipe fingerprint sensor an individual will move his or her finger in a swipe motion over the capacitive sensor. The sensor will capture several slices of the fingerprint as it moves over the sensor and then will stitch it together to form a complete fingerprint (Figure 3.6). Swipe sensors are becoming popular in small-scale devices such as cell phones, PDAs, and laptops as the physical space requirement is smaller and cheaper to manufacture.

3.3.5 RF Sensor

RF sensors are also solid state silicon sensors that use RF signals to measure the contours of the ridges and valleys of the finger skin. A small RF signal is generated by the sensor that travels through the finger, as illustrated in Figure 3.7. The receiving sensors measure the difference in the signal that is dependent on the distance between the finger skin and the capacitive plate. The fingerprint structure is captured at the dermal level, which makes it resistant to epidermal changes.

3.3.6 Thermal Fingerprint Sensor

Thermal fingerprint sensors use thermal energy flux to capture fingerprints. When a ridge is in contact with a sensor surface of a different temperature, heat flows between the ridge and the sensor surface. The sensor surface is made up of an array of micro-heater elements, and a cavity is formed between the valley of the fingerprint surface and the heater element [15]. Since the valley is not in contact with the sensor surface, there is no heat flow between the valley and the sensor surface. The heat flux is measured and converted into a digital representation of the fingerprint

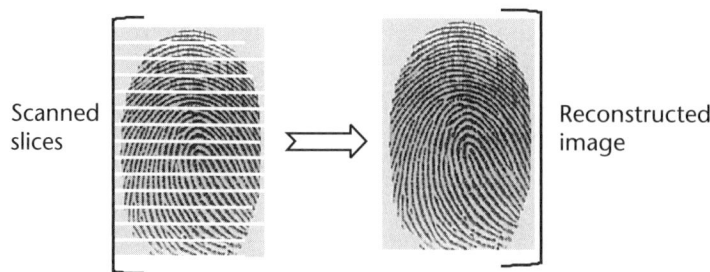

Figure 3.6 Swipe capacitive sensor.

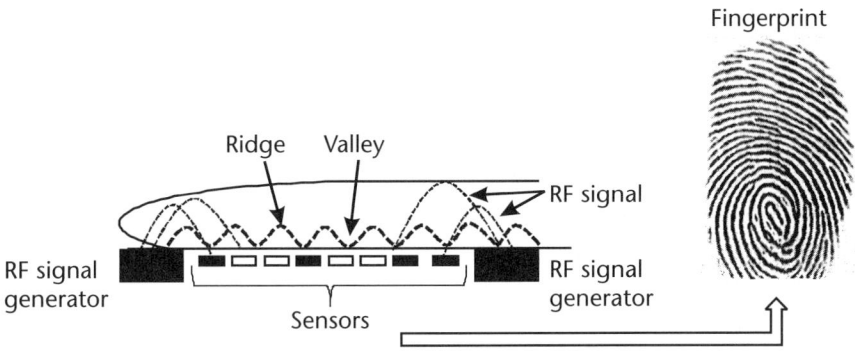

Figure 3.7 RF fingerprint sensor.

surface. Since thermal equilibrium is reached quickly between the finger and the sensing material, a swipe action is required to capture a detailed fingerprint image.

3.3.7 Electro-Optical Sensor

Certain polymers are capable of generating light when appropriate voltage is applied to them. Electro-optical sensors are made of this polymer acting as the platen and a CMOS camera directly underneath it capable of capturing the emitted light. When the ridgelines come in contact with the polymer, the platen emits light and resulting ridge pattern is captured by the CMOS camera.

3.3.8 Multispectral Imaging Sensor

Multispectral imaging is a variation of optical technology that captures images of the finger at several different wavelengths and optical conditions [16]. Complementary details of the finger at the epidermis and dermis layers are captured in the multiple images, thus providing more information of the fingerprint. These images are then fused to form a single fingerprint image.

3.3.9 Other Technologies

Fingerprint sensors based on ultrasonic and piezoelectric technologies have also been created, but their commercial adoption is still relatively low compared to other sensors. Ultrasonic fingerprint sensors use differences in reflected sound waves to recreate ridges and valleys of the finger skin. Piezoelectric sensors measure the difference in pressure between ridges and valleys when the finger skin touches the sensor membrane and converts it into an electric signal. Touchless fingerprint sensors are also available that do not require any physical interaction and thus do not introduce any physical deformation into the resulting fingerprint image.

3.3.10 Limitations of Acquisition Technologies

All of the sensor technologies discussed in this section have advantages and disadvantages that make them more suitable for specific applications than others. Table 3.1 provides a comparative analysis of limitations of various technologies.

Table 3.1 Fingerprint Sensor Limitations

Sensor	Limitations
Optical	Captured fingerprint images are affected by residue of previous fingerprints and ambient light focused on platen.
	Finger skin conditions such as oiliness, moisture content, and elasticity have a significant impact on captured fingerprint images.
Capacitive	Sensors are affected by electrostatic discharge (ESD) and wet fingers in general.
	Production cost increases exponentially with the increase in size of the sensor area.
	Higher cost and longer time are required for training on swipe sensors.
	The silicon layer needs to be protected.
RF	Captured details at the epidermal and dermal levels of skin provide high-quality images from dirty and poor skin quality fingers.
	Sensors are affected by electrostatic discharge (ESD) and wet fingers.
	The production cost increases exponentially with the increase in the size of the sensor area.
	Higher cost and longer time are required for training on swipe sensors.
	The silicon layer needs to be coated.
Thermal	Ambient temperature has an impact on the fingerprint sensor.
	Thermal sensor requires swipe technology and thus has a higher cost of training.
Electro-optical	The skin condition has an impact on the quality of fingerprint images.
Multispectral imaging	It has a relatively larger form factor compared to other sensors and a relatively higher cost.
Ultrasonic	It is comprised of several mechanical parts, results in a larger size, and has a relatively higher cost.
Touchless	It has a relatively larger form factor compared to other sensors and a relatively higher cost.
	Finger skin conditions such as oiliness, moisture content, and elasticity have a significant impact on captured fingerprint images.

3.3.11 Fingerprint Impressions

There are multiple methods of presenting a fingerprint to the sensor (see Figure 3.8). *Slap fingerprints* are multiple flat fingerprints captured at the same time. The segmentation of multiple fingerprints into individual fingerprints is a key challenge for slap fingerprints. NIST conducted an open evaluation of segmentation algorithms in 2004 that is discussed in Section 3.11.5. Before the advent of live scan systems, they were collected on paper cards and stored for future use. *Rolled fingerprints* capture information from one end of the nail to the other. Capturing rolled fingerprints requires a trained operator and is a relatively slow process. Good-quality rolled fingerprints capture approximately twice as much information

Slap fingerprints   Rolled fingerprints   Flat fingerprint

Figure 3.8 Slap fingerprint, rolled fingerprint, and flat fingerprint. (*Source:* NIST Interagency Report 7209, NIST, 2004. Reprinted with permission.)

as slap fingerprints. Both slap and rolled fingerprints can be captured using inked and live scan methods. Traditionally, slap and rolled fingerprints have been used in law enforcement, but their use is expanding to large-scale systems as multiple fingerprints provide a higher level of accuracy. Most commercial systems capture a single fingerprint, which is also called a *single-finger flat print*.

3.4 Fingerprint Feature Extraction

The information contained in a fingerprint can be categorized into three different levels.

- *Level 1:* The global ridge pattern constitutes features at this level. The flow of the pattern, the frequency of ridges, and the orientation of the ridges can be determined at this level. The classes of the Henry classification scheme is an example of a Level 1 feature.
- *Level 2:* Minutiae details of the fingerprint are included in this level. The total number of minutiae points, the type of minutiae, the angle of the minutiae, and the x-, y-coordinates can be determined at this level. A minutiae point, or a Galton feature, can be of several types, as illustrated in Figure 3.2. Most automated fingerprint systems categorize minutiae into ending, bifurcation, or other types, as this has been shown to be sufficient for matching purposes. A more extensive categorization of minutiae types is generally used in the forensic examination of fingerprints.
- *Level 3:* Extremely highly detailed features such as width, shape, and contour of ridges, shape of incipient ridges, and distribution of pores in ridges constitute features at this level. In order to capture these features, 1,000 dots per inch (dpi) sensors are recommended, although there are 500-dpi sensors capable of capturing some of these features.

The fingerprint image also consists of singularities called *core* and *delta,* shown in Figure 3.9. The core is the point of maximum curvature on the innermost ridgeline and is also sometimes referred to as the center of the fingerprint. The delta is the point where two ridgelines moving in parallel change direction and move away from each other. These singularities are often used as anchor points for fingerprint alignment and matching purposes.

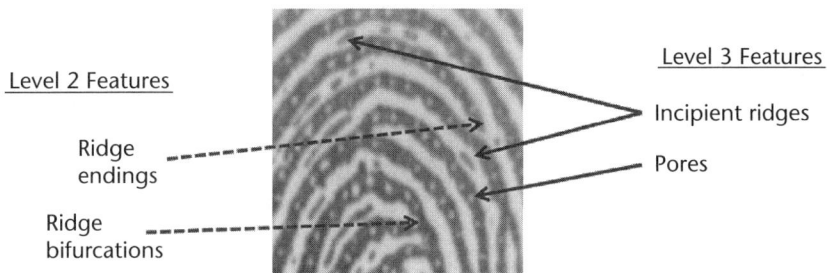

Figure 3.9 Level 2 and Level 3 features.

Fingerprint feature extraction will generally follow these steps: orientation field estimation, segmentation, enhancement, binarization, skeletonization, and minutiae extraction.

The orientation field of a fingerprint image describes the underlying shape of the fingerprint and is an extremely important step of the extraction process. The orientation field contains information about the direction of the ridgelines and is computed for every pixel in the fingerprint image.

Segmentation of the fingerprint image involves identifying the region of interest in a fingerprint and separating it from the background.

A fingerprint image can have spurious information embedded in it, such as cuts, scars, false minutiae points, and other artifacts that degrade the quality of information extracted. In the enhancement stage the contrast between ridgelines and valleys is increased, ridgelines are smoothed out, gaps are filled, and regions of the fingerprint image that cannot be recovered are flagged so that they are not included in the final extraction process.

All of the previous steps, illustrated in Figure 3.10, are performed on a gray level fingerprint image. Now the fingerprint image is converted into a binarized image comprised of only black and white pixels, where the black pixels correspond to ridges and the white pixels correspond to valleys. The ridgelines of a binarized image are eroded until they are only 1 pixel wide, resulting in a skeletonized image. Once a skeletonized image is produced, minutiae extraction is a relatively simple process. Each ridgeline is traced until there is a discontinuity that is marked as a minutiae point and the type of minutiae point is determined using heuristic algorithms. The angle of the minutiae point is calculated as the angle between the ridgeline on which the minutiae is located and the horizontal axis.

It should be noted that each step of the extraction process has been studied in depth by researchers over the last 3 decades and there is a vast body of knowledge related to this area. Readers interested in the algorithmic details should read [17, 18].

3.4.1 Quality Assessment

Fingerprint image quality has a significant impact on performance and several techniques have been devised to measure the quality of fingerprints. These techniques can be categorized into three groups [19]:

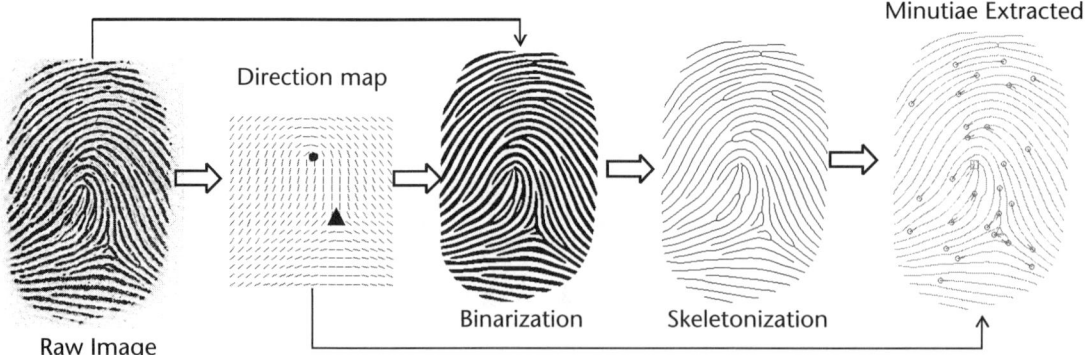

Figure 3.10 Minutiae extraction steps.

1. *Local features:* These methods usually divide each fingerprint image into several nonoverlapping blocks and compute a quality metric for each block. The local block quality measurements are then aggregated to form an overall quality score. Certain methods use a weighting method to give preference to quality scores of blocks close to the core.
2. *Global features:* These methods analyze the overall image and compute a quality score. The analysis of the orientation field and the uniformity of ridgelines are commonly used. Any abrupt disruption in the flow of ridgelines or significant differences in the uniformity of ridge to valley thickness can be viewed as a sign of poor quality. Power spectrum analysis using 2-D discrete Fourier transform yields a region of interest of the spectrum represented as a series of ring-shaped regions. As the quality of the fingerprint increases, the energy of the spectrum will be more concentrated within the ring-shaped regions.
3. *Neural networks:* This method uses quality as a predictor of separation between match and nonmatch scores. National Fingerprint Image Quality (NFIQ) an algorithm created by the National Institute of Standards and Technology (NIST), is one of the most popular examples of this type of quality assessment. NFIQ is a neural network that was trained using a large number of input vectors consisting of 11 fingerprint features such as total number of minutiae, size of foreground, number of minutiae at various quality levels, and quality of different zones of the fingerprint image. The output was mapped to a score between 1 and 5, where 1 represents the best possible score and 5 represents the worst possible score. Readers who are interested in the technical details are encouraged to read [20].

Image quality techniques that use local and global features measure the character and fidelity of the fingerprint image, while NFIQ measures the utility of the fingerprint image. A technical report was published in 2010, ISO/IEC TR 29794-4:2010, which proposed a framework for assessing the quality of fingerprints and interpreting the quality scores in relation to the matching performance [21].

3.5 Fingerprint Compression

Although an extracted fingerprint template is significantly smaller in size than the fingerprint image several deployments might decide to store fingerprint images for future use. Also data collection centers in a distributed architecture might be required to send fingerprint images to a central processing center. These requirements put an enormous strain on storage and transmission requirements thus necessitating robust fingerprint compression techniques. For fingerprint images even the smallest minutiae and ridge line details are important for recognition purposes which need to be preserved during the compression process. The problem is that most image compression techniques remove these high frequency details. To overcome this issue, the FBI created an image compression algorithm called Wavelet Scalar Quantization (WSQ) for the compression of fingerprint images; this algorithm is now the standard used in law enforcement for exchanging compressed images [22]. The current FBI standard specifies a WSQ compression ratio of 15:1 that can be applied to only 500-dpi images. Research conducted by NIST showed that the performance of fingerprint recognition degrades significantly beyond a compression ratio of 20:1. The latest version of the JPEG2000 compression algorithm is also gaining acceptance, and currently the fingerprint image exchange standards at ISO/IEC 19794-2 allow both WSQ and JPEG2000 compression. The ANSI/NIST-ITL standard allows fingerprint images of 1,000 dpi or more to use JPEG2000.

3.6 Fingerprint Feature Matching

A fingerprint captured from the same finger on the same sensor will never be exactly the same. This can be due to one of several reasons: the finger did not make proper contact with the sensor resulting in a partial impression, the finger was not placed with the same force or the same orientation as the previous attempt, or the finger was affected by an external factor such as an abrasion or a scar. A fingerprint matcher has to be capable of matching fingerprints that have significant differences from the same individual. A human examiner will typically match two fingerprints by first analyzing the first level features and then compare the corresponding second level features such as the type of minutiae points and relative distances between minutiae points. Although automated fingerprint matchers follow some of these principles in their implementations, they can take advantage of signal processing techniques unavailable to human examiners. The body of research related to fingerprint matching is vast, but most matchers can be categorized into one of three groups, also summarized in Table 3.2:

1. *Minutiae-based matching:* This is the most popular form of fingerprint matching and most commercially available systems belong to this family. The minutiae extraction process typically provides the x-coordinate, y-coordinate, angle, and type of the minutiae point; these details are used for comparison. In most real-world situations the minutiae sets of the two fingerprints being compared will not have the same number, nor will they have the same rotational orientation. The alignment of these two minutiae sets is

Table 3.2 Characteristics of Fingerprint Matching Algorithms

Minutiae Based	*Correlation Based*	*Ridge Feature Based*
Requires extraction of minutiae details such as x-coordinate, y-coordinate, angle, and type	Requires global level features and robust alignment technique	Requires a division of fingerprint image into several regions and robust alignment technique
Quality of fingerprint image has an impact on minutiae extraction and accuracy	Nonlinear distortions have a significant impact on accuracy	Accuracy is minimally impacted by minutiae points
Requires a large amount of fingerprint image so that enough minutiae points can be extracted	Can operate on a small region of interest of the fingerprint image	Can operate on a small region of interest of the fingerprint image
Computation complexity is low	Computational complexity is high	Computational complexity is high
Works well in identification application	Works well in verification applications	Works well in verification applications

extremely important for robust matching. The Hough transform is a popular alignment technique in image processing, and several algorithms have been proposed based on this technique. Once alignment is completed, the number of overlapping minutiae can be computed and compared against a threshold to generate a decision. Graph-based and distance-based matching algorithms exploit the spatial relationship between minutiae points. These algorithms do not require alignment because, in theory, relative distances between minutiae points remain constant irrespective of fingerprint orientation. Of course, this is not the case in real-world applications, and these algorithms use a prespecified error tolerance parameter to find the level of spatial overlap between the two minutiae datasets.

2. *Correlation-based matching:* This family of algorithms uses the correlation of pixel intensities of superimposed fingerprint images to find a degree of similarity between two images. One of the fingerprint images is rotated at several angles to compensate for rotation variance and compared to the other fingerprint image. This is a type of global matching, as it does not use local minutiae features for matching.

3. *Ridge-based matching:* This family of matching algorithms uses texture information, localized orientation and frequency, and ridge shape features. These algorithms typically extract features from several nonoverlapping regions of the fingerprint. This method can also be applied to only a certain region of interest of a fingerprint and does not require a full fingerprint image. Ridge-based matching has been used in combination with minutiae-based matching to provide enhanced accuracy.

Most automated fingerprint recognition systems are based on Level 1 and Level 2 features, although human examiners do use Level 3 features for matching fingerprints. There is an area of active research in automated Level 3 feature extraction and matching, but the size of fingerprint image data requirements and the high cost of sensors have limited the development of commercial systems based on Level 3 matching. Rather, Level 3 features have attracted a lot of attention from the law enforcement community, and a Committee to Define an Extended Fingerprint Feature Set (CDEFFS) was formed in 2005 to identify Level 3 details that can be used in large-scale automated fingerprint recognition. The findings of this committee are expected to be incorporated into the updated ANSI/NIST-ITL standard, which is the standard used for the interchange of data in the law enforcement community and a reliable indicator of the direction of commercial development.

3.7 Classification and Indexing

Large-scale fingerprint recognition systems are capable of storing fingerprints in excess of tens of millions. For such systems a full database comparison is not an acceptable solution for identification transactions. In order to optimize the matching process, classification and indexing techniques are used to store the fingerprint data. Classification refers to the separation of fingerprint data based on some mutually exclusive characteristics that can be associated with the fingerprints. An example is the classification of fingerprints based on the pattern of the flow of ridgelines such as the Henry classification. Certain systems attach biographical information such as gender or the finger number to the fingerprints for further classification, but such information might not always be available. A problem with the classification based on the ridge flow pattern is the uneven distribution of these patterns, as shown in Table 3.3.

This uneven distribution does not lend itself to an effective classification solution. To overcome this problem, indexing techniques have been researched as well. Indexing does not attempt to partition fingerprints into nonoverlapping classes, but rather represents a fingerprint as a vector of its spatial features. The indexing process entails a series of queries that keeps on returning a successively smaller list of matching fingerprints, and once a small enough list is obtained, fingerprint matching can be performed on the list. Indexing is also called continuous classification because of the nature of the process.

3.8 Synthetic Fingerprint Generation

Any type of a fingerprint system evaluation requires fingerprints from a substantial number of individuals. This can be a costly and time-consuming exercise that needs to be conducted with extreme care to avoid any labeling errors. A solution to this problem is given by the technique of synthetic fingerprint generation. Synthetic fingerprint generation creates computer models of human fingerprint data with enough intraclass and interclass variation to represent differences observed in nature. SFinGe is a synthetic fingerprint generating software and can be an effective tool for the technology testing of fingerprint algorithms [24]. Synthetic fingerprint generation is discussed in Chapter 16.

3.9 Automated Fingerprint Identification System (AFIS)

Fingerprint systems that are used in large-scale implementations are called AFIS. AFIS generally has the following characteristics:

- AFIS is capable of processing inked prints, latent prints, and digital fingerprints from 10-print cards or 10-print scanners.
- It stores tens of millions of fingerprints.

Table 3.3 Distribution of Ridge Flow Patterns

Class	Percentage
Arch	3.70%
Tented arch	33.80%
Left loop	31.70%
Right loop	2.90%
Whorl	27.90%

Source: [23].

- Comparisons occur extremely quickly, using highly sophisticated hardware and indexing/classification techniques.
- Identification generally returns a candidate list.
- Minutiae placement on fingerprint images can be manually annotated or located by experts.

Although AFIS are still predominantly used in law enforcement for criminal identification, their use in government to citizen applications such as welfare disbursement and border control is increasing. Another example is the voluntary Unique Identity Number (UID) being provided by the government of India to all residents. To ensure that a UID is issued to only one resident, an AFIS is being used for the de-duplication of records.

3.10 Standards

Traditionally, the development of standards is influenced by specific applications, and this is also true for fingerprint recognition. The ANSI/NIST-ITL Data Format for the Interchange of Fingerprint, Facial, and Other Biometric Information—Part 1 was originally drafted in 1986 as a means for exchanging fingerprint information between various law enforcement agencies and the FBI. The current version of the standard ANSI/NIST-ITL 2007 incorporates data exchange for fingerprint, face, palm print, iris, scar marks, tattoos, and future modalities. This standard supports seven different types of fingerprint records listed in Table 3.4.

This standard serves as the foundation for three different implementations that are used extensively in the United States and internationally:

1. *FBI EBTS:* The FBI Electronic Biometric Transmission Specification (EBTS) is the FBI's implementation of the standard and is a successor to the Electronic Fingerprint Transmission Specification (EFTS). The FBI formats all its data using this specification, and any agency that wants to match fingerprint data using the FBI system has to format its data according to the FBI EBTS. An important element of the FBI EBTS is Appendix F IAFIS Image Quality Specification, which specifies certification guideline for fingerprint scanners used in collecting data that will be used with the FBI system.

Table 3.4 ANSI/NIST-ITL: 2007 Fingerprint Records

Record Type	Description
Type 3	Low-resolution grayscale fingerprint image
Type 4	High-resolution grayscale fingerprint image
Type 5	Low-resolution binary fingerprint image
Type 6	High-resolution binary fingerprint image
Type 9	Minutiae data that allows for minutiae representation using proprietary methods or ANSI/INCITS 378:2004
Type 13	Latent fingerprint image
Type 14	10-print fingerprint image

2. *DoD EBTS:* This implementation is used by the U.S. Department of Defense to format its fingerprint data for exchange and usage within its systems.
3. *Interpol INT-I:* This implementation is used by all nations that are members of Interpol for the exchange of fingerprint image data.

ANSI/INCITS and ISO/IEC SC 37 have published several U.S. and international fingerprint standards, respectively. These standards provide specifications for exchanging fingerprint information in various formats, including raw image and minutiae data. These standards are discussed in Chapter 12.

NIST published a document titled *Mobile ID Device Best Practice Recommendation Version 1* that provides guidelines for law enforcement agencies for capturing multiple biometric data, including fingerprint, using mobile devices [25]. The best practices define fingerprint acquisition requirements for minimum resolution, minimum dimension, compression algorithm, compression algorithm, sensor certification, minutiae extractor certification, and image quality. The parameters for each capture requirement are mandated according to the intended use of the fingerprint image or template. These guidelines are expected to significantly influence the development of standards for mobile devices.

3.11 Evaluations

Fingerprint recognition technology has undergone several independent evaluations over the course of the last decade. Some of these evaluations have used large datasets, numbering in the millions of fingerprints, collected from operational systems such as border crossing and law enforcement, thus providing a realistic view of system performance. Some of the prominent evaluations are introduced in this section and a detailed discussion of these evaluations, along with publicly available fingerprint databases, can be found in Chapter 13.

3.11.1 Fingerprint Vendor Technology Evaluation (FpVTE)

The Fingerprint Vendor Technology Evaluation (FpVTE) was conducted by NIST to assess identification and verification capabilities of commercially available systems using operational data and determine the effect of extraneous variables on recognition performance. The results of this test were published in 2003 and are

Table 3.5 ANSI/INCITS Fingerprint Standards

Standard	Description
ANSI/INCITS 377-2004	Fingerprint pattern interchange format. It describes the process for converting a raw fingerprint image into a grid-based pattern representation.
ANSI/INCITS 378-2004	Fingerprint minutiae interchange format. It specifies how to determine the placement and describe minutiae points on an image.
ANSI/INCITS 381-2004	Fingerprint image interchange format. It specifies the format, content, and description of the pixel information of raw fingerprint images.

3.11 Evaluations

Table 3.6 ISO/IEC SC 37 Fingerprint Standards and Technical Report

Standard/TR	Description
ISO/IEC 19794-2	Fingerprint minutiae interchange format that specifies placement and description minutiae points on an image. Three types of minutiae are defined: ridge ending, ridge bifurcation, and unknown.
ISO/IEC 19794-3	Fingerprint pattern interchange format that describes the process for converting a raw fingerprint image into two-dimensional spectral representations. The fingerprint image is divided into nonoverlapping cells and each cell is characterized by wavelength in the x and y directions, amplitude, and phase.
ISO/IEC 19794-4	Finger image interchange format that specifies the format, content, and description of the pixel information of raw fingerprint images and acquisition sensor parameters. Two-dimensional grayscale raster images can be uncompressed or compressed using JPEG, JPEG2000, PNG, and WSQ.
ISO/IEC 19794-8	Fingerprint pattern skeletal interchange format that provides interoperability between pattern and minutiae interchange formats.
ISO/IEC TR 29794-4	Image quality standard that specifies the framework for assessing fingerprint image quality and the interpretation of image quality metrics.

publicly available [26]. The FpVTE was conducted as three separate tests: large-scale, medium-scale, and small-scale. Each test was characterized by the size of the enrollment dataset, the type of fingerprint acquisition (rolled, flat, or slap), and the number of fingers associated with each identity. A total of 393,370 distinct fingerprint images were used in this evaluation, and more than 34 systems were evaluated over the different tests. The key observation was that the accuracy of large-scale systems can be increased by improving image quality and number of fingers used for that same individual. The results showed that the most accurate fingerprint system generated a 99.4% true accept rate at a 0.01% false accept rate and a 99.9% true accept rate at a 1.0% false accept rate for single fingerprint images. When four fingerprint images were used for the recognition of a single individual, the true accept rates were higher than 99.9% at a 0.01% false accept rate [26]. This test highlighted a variety of operational factors that have a clear impact on recognition performance and identified focus areas for future research.

3.11.2 Minutiae Interoperability Exchange (MINEX)

NIST initiated a program called the Minutiae Interoperability Exchange (MINEX) that is aimed at evaluating the performance of ANSI/INCITS 378 templates for various types of applications. The MINEX 2004 test evaluated the capability of various fingerprint vendors to generate and match ANSI/INCITS 378:2004 templates. FNMR was calculated for various test scenarios at a fixed FMR of 1%. The results showed that proprietary templates performed better than the standardized templates and two-finger performance using standardized templates was better than single-finger proprietary templates. The complete results are publicly available [27].

MINEX II, another evaluation program under the MINEX umbrella, was initiated by NIST to evaluate the accuracy and speed of fingerprint minutiae matchers operating on ISO/IEC 7816 smartcards. The minutiae templates are stored on the card in the ISO/IEC 19794-2 Compact Card format. Phase IV of the MINEX

II series evaluated the accuracy and speed of single-finger and two-finger fusion matching of 17 different systems using four different combinations of data sources. The fastest implementation had an execution time for genuine comparisons of 0.076 second, and the slowest implementation had an execution time of 1.173 seconds. The results for the most operationally viable scenario are summarized in Table 3.7. These results are publicly available [28].

Based on the MINEX 2004 evaluation, NIST has established the MINEX certification that is given to fingerprint vendors who can successfully generate ANSI/INCITS 378 templates and match them against templates generated by other certified vendors. The list of certified feature extractors and matchers are also considered Personal Identity Verification (PIV) compliant [29]. The PIV card contains biometric data of its owner and is used for the verification of federal employees and contractors. PIV is discussed in detail in Section 3.12.2.

3.11.3 Minutiae Template Interoperability Test (MTIT)

The MTIT project is funded by the European Commission under the Sixth Framework Programme with the objective of testing the interoperability of ISO/IEC 19794-2 minutiae templates. The MTIT follows a similar testing protocol as MINEX. Other objectives of the MTIT are to define criteria for interoperability testing, develop a testbed for interoperability testing, and deliver a certification process for the interoperability of fingerprint generators and matcher.

3.11.4 Fingerprint Verification Competition (FVC)

The Fingerprint Verification Competition (FVC) is a technology evaluation conducted by a collection of universities. The FVC was started in 2000 as an open competition to judge which fingerprint extraction and matching algorithm performed the most accurately and reliably on a fingerprint dataset. The FVC was conducted in 2000, 2002, 2004, and 2006, and each competition had live fingerprint data collected on different types of sensors as well as synthetic fingerprints. The FVC has now evolved into a new initiative called *FVC-onGoing*. FVC-onGoing is a Web-based evaluation of fingerprint extractors and matchers on a set of four different datasets that include raw fingerprint images, challenging raw fingerprint images, 19794-2 templates, and challenging 19794-2 templates. Participants can upload their extractors and matchers and receive performance metrics on these databases. Results from the FVCs are summarized in Table 3.8 and more detailed information can be found on the FVC Web site [30].

Table 3.7 Summary of MINEX II: Phase IV

	Lowest FNMR %	Highest FNMR %	Average FNMR %
Single finger at FMR = 0.05%	3.25	15.37	6.19
Two-finger fusion at FMR = 0.1%	0.42	5.50	1.28

Source: [28].

3.11 Evaluations

Table 3.8 Summary of Fingerprint Verification Competitions (FVC)

Year	Number of Algorithms	Sensor Details	Lowest EER (%)
2000	25	KeyTronic: Optical Sensor	0.67
		ST Microelectronics: TouchChip	0.61
		Identicator Technology: DF90	3.64
		Synthetic Fingerprint Generator	1.99
2002	33	Identix: TouchView II	0.10
		Biometrika: FX2000	0.14
		Precise Biometrics: 100SC	0.37
		Synthetic Fingerprint Generator	0.10
2004	67	CrossMatch: V300	1.97 (Open) 3.89 (Light)
		DigitalPersona: U.are.U4000	1.58 (Open) 4.01 (Light)
		Atmel: FingerChip FCD4B14CB	1.18 (Open) 2.92 (Light)
		Synthetic Fingerprint Generator	0.61 (Open) 1.88 (Light)
2006	70	Electric field sensor	5.56 (Open) 5.35 (Light)
		Optical sensor	0.02 (Open) 0.14 (Light)
		Thermal Sweeping Sensor	1.53 (Open) 1.63 (Light)
		Synthetic Fingerprint Generator	0.26 (Open) 0.42 (Light)

Source: [30].

It should be noted that the results from the different FVCs should not be compared as the data was collected using various sensors and data collection protocols.

3.11.5 Slap Fingerprint Segmentation Evaluations

Slap fingerprints collected using live scan systems and fingerprint cards have to correctly individualize fingerprints and additionally can exhibit several problems, including:

- Fingerprints captured at an angle requiring a rotation of up to 45°;
- Missing fingers;
- Notes written on them in case of fingerprint cards;
- Overlap between fingers and thumbs;
- Background noise.

In 2004 NIST conducted the SlapSeg04 evaluation to assess the accuracy of segmentation algorithms. In 2009 NIST conducted a follow-up evaluation called SlapSegII that used operational data collected on 2-inch and 3-inch platen devices. SlapSegII used manually created bounding boxes for assessing the accuracy of the algorithms, whereas SlapSeg04 used automated matching to assess the accuracy. In

SlapSegII the majority of the algorithms were capable of segmenting at least three fingers from a single four-print slap impression for 96–99% of the impressions. The evaluation also indicated that two-thumb impressions captured on 3-inch platen proved to be more difficult to segment than four-print impressions due to a lack of relative difference in height between adjacent impressions and the rotation of thumbs in the opposite direction. In four-print impressions all the fingerprints are rotated in the same direction and there is a relative difference in height between adjacent prints, which aids in the segmentation process. Complete results for this study are publicly available [31].

3.12 Applications and Trends

This section discusses a few prominent operational fingerprint-based systems.

3.12.1 Integrated Automated Fingerprint Identification System (IAFIS)

IAFIS, started in 1999, is a U.S. national level fingerprint and criminal history database operated by the FBI, which also assists local, state, and federal agencies in identifying criminals and terrorists. Along with fingerprints, the IAFIS also contains scars, tattoo marks, mug shots, weight, height, and other physical characteristics. The IAFIS currently stores approximately 66 million criminal records and 25 million civil records and is capable of responding to a criminal record submission within 10 minutes and a civilian record submission within 24 hours. The Next Generation Identification (NGI) system is going to replace the current IAFIS in the near future.

3.12.2 Personal Identity Verification (PIV)

The Homeland Security Presidential Directive (HSPD) 12 mandated a common identification standard for federal employees and contractors. In response to HSPD 12, NIST developed the Federal Information Processing Standard (FIPS) 201, which specifies a Personal Identity Verification (PIV) card for federal employees and contractors; this standard uses fingerprints and face data stored on a secure card that operates both in contact and contactless modes. Before a PIV card is issued, a background check is carried out using all 10 fingerprints out of which two fingerprint templates and the face image are stored on the secure card. A PIV card is issued that contains physically viewable and electronically stored data that is used for physical and logical access. During verification the authorized owner provides a live fingerprint sample, which is compared against the template stored on the card. The NIST SP 800-73 provides the PIV data elements, identifiers, structure, and format and describes the programming and card interfaces and requirements that enable the PIV identity credentials to be used interchangeably throughout federal agencies.

3.12.3 US-VISIT

The United States Visitor and Immigrant Status Indicator Technology (US-VISIT) program, launched in 2004, collects fingerprint data and face images from visitors entering the United States. This system is called the Automated Biometric Identification System (IDENT) and currently contains more than 90 million sets of prints and is growing every day [32]. Originally, this program collected two fingerprints and was then expanded to collect all 10 fingerprints for improved accuracy. The objective of the program is to ensure that visitors entering the United States are correctly identified, known terrorists and immigration violators cannot enter the United States, and the person's biometrics match the identification document presented to the border control officer.

3.13 Design and Deployment Considerations

For applied researchers and practitioners, it is extremely important to view performance as a function of the user, the fingerprint subsystems, and operational requirements. These requirements are discussed next.

3.13.1 User Interaction

User interaction is one of the most overlooked aspects of biometric systems in general. For fingerprint recognition systems, improving user interaction by providing training is an effective means of improving performance of the system. Fingerprint sensors can be of the touch or swipe interaction type (swipe interaction has a higher learning curve). Capturing consistent images from touch fingerprint sensors with a smaller capture area is more difficult than sensors with a larger capture area. These issues can be addressed using proper user training. For unsupervised fingerprint systems, training is an effective means of improving user experience and capturing high-quality fingerprints. An illustration of poor-quality fingerprints is shown in Figure 3.11.

3.13.2 Environment Issues

Environmental conditions can have a significant impact on the sensor acquisition and fingerprint skin conditions, thereby impacting performance of the system. Direct illumination on the platen of an optical fingerprint sensor interferes with the capture system and results in poor-quality images. Thermal fingerprint sensors require a controlled ambient temperature for the sensor to be able to capture fingerprints. Researchers collected fingerprints on optical and capacitive fingerprints over the course of 1 year where the temperature range varied from −30°C to +20°C [33]. The study found that there was no significant difference in image quality and performance between the two sensors due to temperature and humidity. However, the researchers did list some technical issues that were a result of the fluctuation in weather. The optical sensors showed signs of condensation underneath the platen on wet and snowy days. The capacitive sensor became too hot under direct sunlight for appropriate usage. The study showed that fingerprint capture sensors are affected more by the temperature change, rather than the skin characteristics of the users.

3.13.3 User Demographics

Will the users of the system exhibit specific characteristics that affect the performance of the fingerprint system? For example, research has shown that fingerprints from the elderly population are more challenging for fingerprint recognition than fingerprints from the younger population [34]. Age and occupation are shown to have an impact on performance, and necessary countermeasures should be taken to reduce their adverse impact on performance. Certain medical conditions prevent users from providing good-quality fingerprint images. Arthritis and Parkinson's disease directly impact the ability to physically interact with the sensor. Individuals affected by the rare genetic disorder called dermatopathia pigmentosa reticularis (DPR) do not have ridge structures on their finger skin and thus cannot provide fingerprints [35].

3.13.4 Interoperability

Fingerprint systems that are designed as a distributed architecture or that need to exchange data with other systems have to tackle the issue of interoperability. Technology interoperability issues can exist at the sensor, feature extraction, and feature matching subsystems. Research has shown that fingerprint images collected from the same individual on different sensors have a higher probability of generating matching errors [36]. NIST has conducted several tests under the Minutiae Interoperability Test Exchange (MINEX) program since 2004 to study the impact of minutiae extractors and minutiae interoperability on matching performance. From a design perspective, a simple solution is to use the same sensor, extractor, and matcher for the entire system, but today's distributed systems make such a design choice impractical.

3.13.5 Spoofing Attacks

Spoofing attacks are becoming more prevalent as individuals attempt to circumvent fingerprint recognition systems. These include methods such as creating fake fingers and surgically modifying fingerprints to influence the recognition process. A detailed discussion of various spoofing methods, countermeasures, and open evaluations of antispoofing algorithms, such as the LiveDet competition, is provided in Chapter 15.

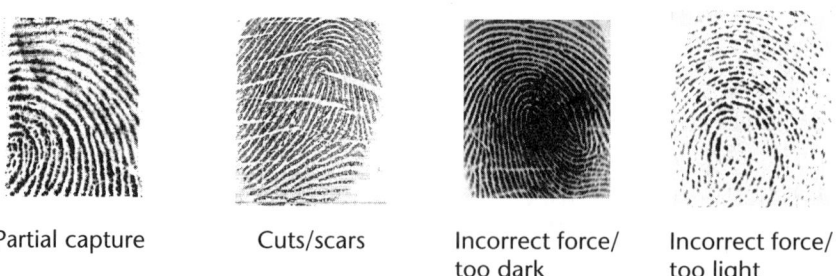

Partial capture   Cuts/scars   Incorrect force/too dark   Incorrect force/too light

Figure 3.11   Poor quality fingerprint samples.

3.13.6 Large-Scale Systems

Identification systems built for extremely large populations such as national ID programs have a completely different set of design level issues to address. For such systems to operate at a high level of accuracy, they have to use more than one finger from the user for matching purposes. Efficient retrieval of data from large databases requires classification, indexing, or a combination of both to complete the operation in a short time. Managing and maintaining large-scale fingerprint databases of raw images can be a time-consuming and financially expensive task. Systems that require the storage of raw fingerprint images need to use an appropriate compression algorithm and an acceptable compression ratio based on feasibility studies. Secure storage of the fingerprint data is the paramount concern. Biometric data should be treated as personally identifiable information and commensurate security measures need to be followed to keep the data confidential and safe from theft.

3.14 Summary

Fingerprint recognition is the oldest and most widely adopted biometric technology, but, as discussed in this chapter, it is by no means a fully mature technology. The improvement of fingerprint recognition requires research into issues that arise from real-world deployments, such as user interaction and system security and policies, along with image processing algorithms. The increase in the use of mobile and small-scale devices for fingerprint recognition is the next frontier. This will introduce a variety of challenges including user interaction, quality assessment in the field, remote connectivity, and policies and procedures to support a mobile infrastructure. Fingerprint recognition still is relatively cheaper than most other biometric solutions and will continue to enjoy broad acceptance in commerical and government implementations. The success of fingerprint recognition in operational deployments will depend on creating solutions that include the user, the system, and the organizational policies.

References

[1] Babler, W. J., "Embryologic Development of Epidermal Ridges and Their Configurations," *Birth Defects, Original Article Series*, Vol. 2, 1991, pp. 95–112.

[2] Prabhakar, S., A. Jain, and S. Pankanti, "On the Individuality of Fingerprints," *IEEE Transactions on Pattern Analysis and Machine Intelligence*, Vol. 24, August 2002, pp. 1010–1025.

[3] O'Gorman, L., and X. Xia, "Innovations in Fingerprint Capture Devices," *Pattern Recognition*, Vol. 36, January 2001, pp. 361–369.

[4] Cole, S. A., *Suspect Identities: A History of Fingerprinting and Criminal Identification*, Cambridge, MA: Harvard University Press, 2001.

[5] Hutchings, P. J., "Modern Forensics: Photography and Other Suspects," *Cardozo Studies in Law and Literature*, Vol. 9, 1997, pp. 229–243.

[6] Galton, F., *Finger Prints*, London, U.K.: MacMillan, 1892.

[7] Bolle, R., S. A. Cole, and N. Ratha, *History of Fingerprint Pattern Recognition*, New York: Springer, 2004, pp. 1–25.

[8] Haylock, S. E., "Khan Bahadur Azizul Haque," *Fingerprint World*, Vol. 5, July 1979, pp. 28–29.

[9] Edward, H., *Classification and Uses of Finger Prints*, London, U.K.: George Routledge & Sons Ltd., 1901, p. 123.

[10] IBG, *The Henry Classification System*, New York: 2003.

[11] Reed, B., "Automated Fingerprint Identification: From Will West to Minnesota Nine-Fingers and Beyond," *Journal of Police Science and Administration*, Vol. 9, 1981, pp. 318–319.

[12] Prabhakar, S., et al., *Handbook of Fingerprint Recognition*, 2nd ed., New York: Springer, 2003.

[13] Bolle, R., N. Ratha, and D. R. Setlak, *Advances in Fingerprint Sensors Using RF Imaging Techniques*, New York: Springer-Verlag, 2004, pp. 27–53.

[14] Lim, T. W., and M. Moghavvemi, "Capacitive Fingerprint Sensor Chip for Automatic Matching," *Proceedings of TENCON 2000*, Vol. 2, Kuala Lumpur, Malaysia, 2000, pp. 442–446.

[15] Han, J., et al., "Thermal Analysis of Fingerprint Sensor Having a Microheater Array," *1999 International Symposium on Micromechatronics and Human Science*, Nagoya, Japan, 1999, pp. 199–205.

[16] Rowe, R. K., K. A. Nixon, and S. P. Corcoran, "Multispectral Fingerprint Biometrics," *6th Systems, Man, and Cybernetics Information Assurance Workshop*, 2005, pp. 14–20.

[17] Bolle, R., et al., *Feature Extraction in Fingerprint Images*, New York: Springer-Verlag, 2004, pp. 145–180.

[18] Maltoni, D., and D. Maio, "Direct Gray-Scale Minutiae Detection in Fingerprints," *IEEE Transactions on Pattern Analysis and Machine Intelligence*, Vol. 19, January 1997, pp. 27–40.

[19] Aguilar, J., F. Fernandez, and J. Garcia, "A Review of Schemes for Fingerprint Image Quality Computation," *COST 275, Biometrics Based Recognition of People over the Internet*, October 2005.

[20] Tabassi, E., and C. L. Wilson, "A Novel Approach to Fingerprint Image Quality," *IEEE International Conference on Image Processing, 2005 (ICIP 2005)*, Genoa, Italy, 2005, pp. 37–40.

[21] ISO/IEC, *ISO/IEC 29794-1:2009 Information Technology—Biometric Sample Quality—Part 1: Framework*, Geneva, Switzerland, 2009.

[22] CJIS, *WSQ Gray-Scale Fingerprint Image Compression Specification*, 1997.

[23] Wilson, C., G. Candela, and C. Watson, "Neural Network Fingerprint Classification," *Journal of Artificial Neural Networks*, Vol. 1, 1993, pp. 203–228.

[24] Cappelli, R., D. Maio, and D. Maltoni, "Synthetic Fingerprint-Database Generation," *Proc. 16th International Conf. on Pattern Recognition*, 2002, pp. 744–747.

[25] Orandi, S., and R. M. McCabe, *Mobile ID Device Best Practice Recommendation Version 1*, Gaithersburg, MD: National Institute of Standards and Technology, 2009.

[26] Wilson, C., et al., *Fingerprint Vendor Technology Evaluation 2003: Summary of Results and Analysis Report*, Gaithersburg, MD, 2003.

[27] Tabassi, E., et al., *MINEX Performance and Interoperability of the INCITS 378 Fingerprint Template*, Gaithersburg, MD, 2006.

[28] Grother, P. J., et al., *MINEX II Performance of Fingerprint Match-on-Card Algorithms: Phase IV Report*, Gaithersburg, MD, 2011.

[29] NIST, "MINEX Compliant List," 2011.

[30] "FVC-onGoing," 2011, https://biolab.csr.unibo.it/fvcongoing/UI/Form/IJCB2011.aspx.

[31] Watson, C. I., *Slap Fingerprint Segmentation Evaluation II: Procedures and Results*, NIST Technical Report, Gaithersburg, MD, 2009.

[32] EPIC, "EPIC—United States Visitor and Immigrant Status Indicator Technology (US-VISIT)," 2011, http://epic.org/privacy/us-visit.

[33] Stewart, R. F., M. Estevao, and A. Adler, "Fingerprint Recognition Performance in Rugged Outdoors and Cold Weather Conditions," *BTAS'09 Proceedings of the 3rd IEEE international Conference on Biometrics: Theory, Applications and Systems*, Piscataway, NJ: IEEE Press, 2009, pp. 300–305.

[34] Elliott, S. J., and S. K. Modi, "Impact of Image Quality on Performance: Comparison of Young and Elderly Fingerprints," *RASC 2006*, K. Sirlantzis, (ed.), Canterbury, U.K., 2006, pp. 449–454.

[35] Handwerk, B., "Born Without Fingerprints: Scientists Solve Mystery of Rare Disorder," *National Geographic News*, 2006.

[36] Elliott, S. J., S. Modi, and H. Kim, "Statistical Analysis of Fingerprint Sensor Interoperability Performance," *Proceedings of the 3rd IEEE International Conference on Biometrics: Theory, Applications and Systems*, 2009, pp. 294–299.

CHAPTER 4

Face Recognition

Face recognition as an automated technique is relatively new, but humans implicitly use their visual and cognitive capabilities to recognize individuals. The ability of humans to recognize objects is extraordinary, and efforts have been ongoing for more than 50 years to replicate this capability using automated techniques. Advances in computer vision techniques and computing capabilities have resulted in remarkable progress in automated face recognition, primarily for use in law enforcement and criminalistics. The very first applications in face recognition were designed to assist human experts by providing candidate lists of suspects, and their scope has increased as technology capabilities have become more sophisticated. Face recognition can be deployed in a wide range of applications such as surveillance of individuals to computer logon using images captured either from still camera or video streams. Add to this our natural understanding of how to interact with the cameras, and face recognition becomes an extremely appealing technology. The earlier techniques of automated face recognition concentrated on 2-D images, but 3-D techniques are now garnering increased interest. Three-dimensional face recognition techniques can operate in less constrained environments and overcome some of the operational limitations of 2-D techniques. This chapter will discuss 2-D and 3-D face recognition techniques, their advantages and drawbacks, operational considerations, and future applications trends.

4.1 History

The earliest work in 2-D face recognition is that of Woodrow Bledsoe and dates back to the mid-1960s [1, 2]. His team focused on creating computer models to recognize faces and, as a result, designed a semiautomated system in which a human operator located distinguishing features such as eyes, lips, and cheeks on the face and input the coordinates into a computer. A complete table of features and distance between them was generated, which could then be used for classification using pattern matching algorithms. Due to the manual nature of this system, it did not have to be concerned with the pose of the individual or illumination and background artifacts. Work based on this approach continued through the 1970s and 1980s with advances focused on automated detection of facial features. An alternative approach to face recognition was initiated by Kirby and Sirovich in the late 1980s [3]. They applied the mathematical decomposition technique of principal component analysis (PCA) to face recognition and through experiments observed that fewer than 100 uncorrelated variables could be used for describing face images. Turk and Pentland proposed a technique, what is also popularly known as *eigenfaces*, for feature extraction and matching of faces based on the work of Kirby and Sirovich [4]. This seminal work demonstrated a practical real-time face recognition system and spurred on a significant amount of work in this area.

The evolution of automated face recognition has continued with advances in 3-D capture and matching techniques. The earliest work in 3-D face recognition used face morphing models that reconstructed a 3-D model of the face using images from multiple views like frontal and profile [5]. With the advent of new sensing technologies, the capture of 3-D data in real time has become possible and a growing body of research is examining techniques for the efficient use of contour and shape information.

4.2 2-D Face Recognition

4.2.1 2-D Face Image Acquisition, Detection, and Segmentation

A face image acquisition device captures the image as a projection of a 3-D object onto a 2-D plane. These devices can be passive, which uses the ambient light reflectance of the object, or active, which uses the reflectance of light emitted by a dedicated source. Passive devices are the most heavily used and include photographic cameras, Web cams, cameras embedded in mobile devices, and video recorders. Most active devices operate beyond the visible spectrum, typically in the infrared spectrum, although cameras that use a flash and operate in the visible spectrum could also be classified in this category. Commercial systems predominantly use passive devices. Section 4.5 discusses face recognition beyond the visible spectrum. The quality of input images is dependent on the type of camera used, which is currently the biggest differentiator. Web cameras are much cheaper, but also provide lower-quality images compared to a high-resolution digital camera. The various factors affecting face quality are discussed in Section 4.2.2.

The very first step after acquiring an image is to detect the face. A successful face detection approach has to differentiate the face region from a background that can be plain or quite complex, for example, in a tourist place. At a fundamental level the face detection process produces a binary decision; the face is detected successfully or not. Neural networks trained on a large set of uncorrelated images have proven to be quite successful at face detection because of the following properties: the face structure is quite similar across humans and quite different from other objects. Techniques that try to identify face landmarks, specifically the eye region, have also been shown to be robust. Skin tone also has enough discrimination power for it to be used in segmentation of face images. The issue in face detection arises from images where the face is rotated sideways; the eyes are covered with sunglasses or other obstructions. After the successful detection of the face, the next steps involve normalization and segmentation of the face image. Spatial

4.2 2-D Face Recognition

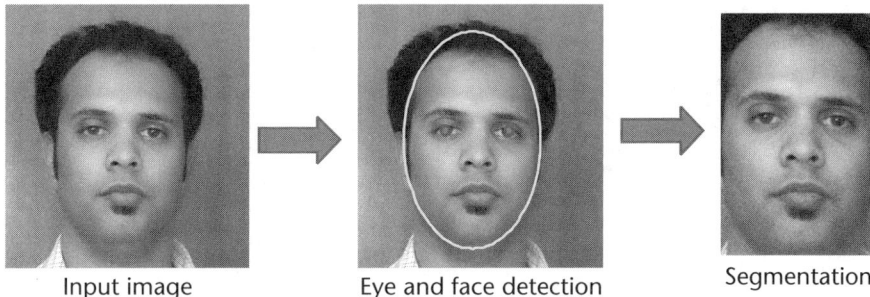

Figure 4.1 Face image detection and segmentation.

normalization is a process that ensures that the pixels between the eyes conform to a predefined measurement and that the face is aligned along the x- and y-axes. The segmentation phase crops the face region from the rest of the image so that the extraction of features becomes a less complex process, shown in Figure 4.1. The quality of the input images also needs to be determined before the feature extraction process is executed, which is discussed in the next section.

4.2.2 Quality Assessment

The efficacy of feature extraction and matching is completely dependent on the correct and repeatable detection of face features. Face image quality assessment is supposed to serve as a predictor of the matching performance of the system. Face image capture can operate in a relatively unconstrained manner, which provides flexibility but also increases the number of variables that can impact the quality of a face image. The face image data interchange standard, ISO/IEC 19794-5, has described several covariates that can affect face image quality (see Table 4.1) along with recommended constraints for improving quality [6]. Some of these problem areas are illustrated in Figure 4.2.

Table 4.1 Face Quality Covariates

Category	Covariate	Constraint
Scene	Pose	Full frontal image
	Illumination	Uniformly spread with no shadow
	Background	Uniformly colored
	Eyes	Open
	Glasses	No reflection with eyes clearly visible
	Mouth	Neutral expression
Photographic	Head position	Centered along vertical axis
	Distance to camera	Face covers more than 50% of image
	Exposure	Appropriate color distribution; gradations of skin tone visible
Digital	Focus	Nonblurred with sharp edges
	Resolution	Greater than 1-mm resolution on face features

 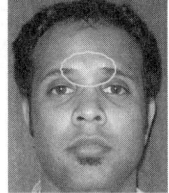  

Incorrect pose and occlusion   Flashlight hotspot   Background clutter   Not full frontal

Figure 4.2 Problematic face images.

The area of face image quality is extremely active, with a substantial body of literature. Empirical results in face quality research has shown that illumination, pose, and camera focus are the most significant quality covariates and that these should be prioritized for quality control purposes. ISO/IEC SC 37 released a technical report in 2010, 29794-5, which defines and specifies methodologies for objective computation of face image quality [7]. This report uses the foundations laid by ISO/IEC 19794-5 and aims to create a common framework for assessing face image quality, which can assist in normalizing face quality scores from multiple algorithms [7]. Figure 4.2 illustrates examples of problematic images for face recognition.

Quality itself is a subjective interpretation, and research has examined differences in quality assessment performed by automated methods and human experts. By identifying factors used by humans for quality assessment, automated techniques can be modeled to replicate human assessment framework, and this is an area of research that be heavily pursued in the near future.

4.2.3 2-D Face Feature Extraction and Matching

Many approaches to 2-D face recognition have been examined that can be generally categorized into appearance-based and feature-based techniques.

Appearance-based methods use global properties of the face image and generally operate on the pixel intensity of the image. The landmarks of the face are not used in describing the face, but rather for preprocessing steps such as alignment and normalization. The most popular appearance-based method, first suggested by Kirby and Sirovich and then refined by Turk and Pentland, uses eigenvalues and principal component analysis (PCA) to extract low-dimensional data from a face image. In this method eigenvectors, or eigenfaces, are calculated from the face image covariance matrix. Eigenfaces essentially represent characteristic features of the face, and a collection of eigenfaces can be used to reconstitute the face image. An image of N pixels will generate N eigenvectors; this is not efficient either computationally or for storage. PCA is used here to identify a relatively small set of vectors that can be used to approximate the original face image. PCA is effective in reducing the dimensionality of the data because face images share a similar structure that is quite different from other 2-D representations of objects. After PCA is applied to the eigenfaces, each face in the dataset can be represented as a weighted

4.2 2-D Face Recognition

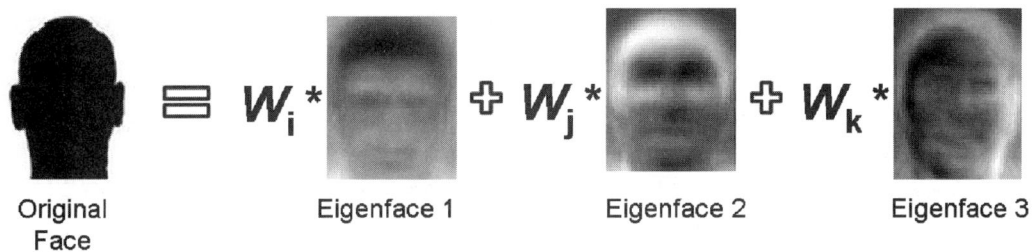

Original   Eigenface 1   Eigenface 2   Eigenface 3
Face

Figure 4.3 Eigenfaces. (*Source:* ORL Face Database, AT&T Laboratories, Cambridge. Reprinted with permission.)

sum of the eigenfaces. These weights are stored as the feature vector for each corresponding face image. The PCA technique can reduce the amount of data required to represent each face by a factor of 1,000. When an image needs to be verified or identified, its vector of weights is calculated using the eigenfaces technique. The vector of weights from the enrolled database that is closest to the new weight vector is declared as the matching face image.

Feature-based approaches use structural comparison of facial landmarks such as endpoints of the eyes and eyebrows, points along contours of the lips, and the nose, as shown in Figure 4.4. Feature-based face matching techniques use geometric properties of the face in combination with graph and distance-based algorithms to compare the spatial relationship between two images. The elastic graph bunch matching (EBGM) technique uses a dynamic link structure based on image graphs to represent a face [8]. The image graph consists of nodes that are located at various face landmarks. These landmarks are extracted in multiple images of the same person, and complex Gabor wavelet coefficients are computed for each node. During recognition the Gabor wavelet coefficients are computed for the test image and compared to the coefficients of images from the enrolled database. This method relies on consistent identification of landmarks and is robust to change in illumination, rotation, and scaling of faces.

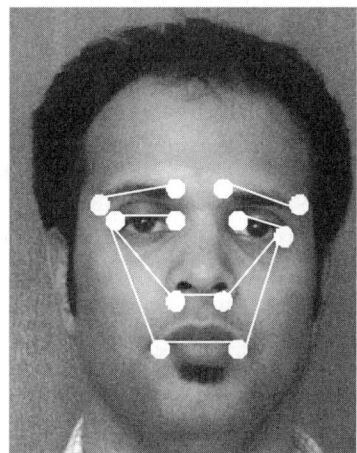

Figure 4.4 Face landmarks.

There are several other approaches that have been examined for face recognition. Linear discriminant analysis (LDA) is a statistical technique for classifying unknown samples into predetermined classes created from a known training set. For the purposes of face recognition, the goal here is to maximize the differences between face images of different people and to minimize differences between face images of the same person. The robustness of LDA is related to the number of training images available, both for the same person and different persons, which is typically limited for real-world applications. To overcome this issue, a hybrid technique was suggested that first performed PCA and then applied LDA [9]. Independent component analysis (ICA), which generalizes the idea of PCA to higher-order statistics, has been examined for facial recognition [10]. ICA is used to take advantage of the information that may be contained in higher-order statistics of image pixels. Line edge map (LEM) uses lines from the face contours as features for matching [11]. This technique tries to replicate the human cognitive process of recognizing objects based only on their outlines, and it has been implemented and tested with varying levels of success. Readers interested in details of these algorithms are encouraged to read [12].

4.2.4 Video Surveillance

Video surveillance is an active area of research within face recognition, especially among law enforcement and intelligence agencies. The ability to identify and track an individual on the move in a complex and dynamic environment is a challenging task. Faces in a video sequence are affected by occlusion, illumination variance, pose variation, and face orientation. There could be multiple faces present in the same frame, which increases the complexity of face matching, as all faces have to be detected and the face of interest has to be segmented. The 2-D face recognition is generally done covertly, although overt systems are used in monitoring open spaces such as shopping malls. The 2-D face surveillance systems have to first address face detection and motion detection, which involves segmenting a moving object, in this case a face, from the rest of the background. The 2-D face surveillance systems have to detect a face region from other objects in the scene. Object tracking is the next step of the surveillance system where the face is tracked from one frame to the next. This involves matching the region of interest over a series of video frames. Motion detection and object tracking work in a cyclical process over all the frames that make up the video sequence. Surveillance systems may also be comprised of multiple cameras, which require analyzing information from multiple sources and deciding which provides the highest quality data. It is unlikely that a full-frontal face image will be available in a single frame. Face information may have to be consolidated from several cameras or frames to create an input sample that can be used for feature extraction. Once a face template has been created, the matching process is similar to still 2-D face images. Commercial systems based on 2-D face surveillance are available on the market.

4.2.5 Beyond the Visible Spectrum

A change in expression or pose changes the 2-D representation of the face significantly, which is directly captured in visible light images. Cameras operating in the visible spectrum cannot capture images in low illumination. To overcome these issues, research has begun on facial recognition for images captured in the infrared (IR) spectrum. IR images have reflectance characteristics that are affected to a lesser degree by illumination, expression and pose, and background noise. Also, the sensitivity of the sensor ensures that reflection from the face is contributing to the majority of intensity variation in the image. Existing feature extraction and recognition techniques have been applied to IR images successfully, although large-scale evaluations have yet to be conducted.

4.3 3-D Face Recognition

Several factors such as pose, illumination, and expression have an impact on pixel intensities of 2-D face images that degrades recognition performance. A 3-D representation of the face that captures the contours and the entire geometry of the face can reduce the effect of environment and expression on matching performance. The following sections will discuss acquisition, extraction, and matching techniques of 3-D face recognition.

4.3.1 Face Acquisition

A 3-D face image is commonly represented using range images or a set of polygons. A range image is an intermediate representation between 2-D and 3-D. It captures depth information of every pixel instead of the intensity of reflected light, as shown in Figure 4.5. Using multiple range images, a polygonal representation of the face can be created. A number of different techniques exist to capture this information. *Stereo imaging systems* use multiple cameras to capture images from different perspectives of the face. Using triangulation techniques in combination with the range information, a 3-D model of the face is created. This type of system was among the first to be commercially produced, but it is affected by the distance between the subject and the cameras as well as environment effects. *Structured light systems* project

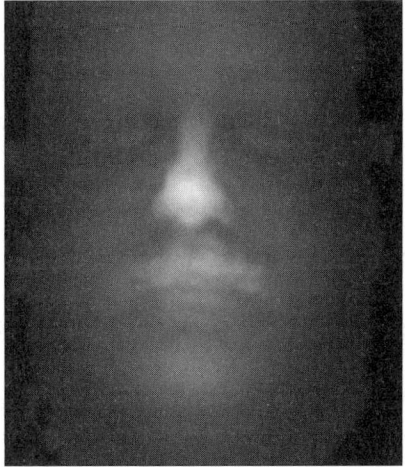

Figure 4.5   Range image. (*Source:* E. Cabello, Rey Juan Carlos University. Reprinted with permission.)

a structured pattern, such as a square or rectangular grid, on the face and measure the distortion to determine depth information. These systems are computationally expensive and accurate measurement is necessary to recreate a model with high fidelity. *Laser systems* scan the face using a laser light while a sensor captures 3-D data of every point of incidence of the light on the face. Once the scan is completed, the 3-D face image is created from collating the scan information. Although highly accurate, currently laser systems are expensive and have high computational processing requirements.

4.3.2 3-D Face Modeling and Matching

Multiple approaches to 3-D face recognition have been examined that can generally be categorized into morphable models and 3-D shape data. Morphable models use a full-frontal 2-D face image that is then fitted to a 3-D model to compensate for pose, illumination, and other variables [13]. The generic 3-D model is representative of the general face structure and, based on the landmarks extracted from the 2-D image, is morphed to create a 3-D model of the face. Once a 3-D model is created, it is compared against the enrolled database and a distance measurement between two models is computed. This method demonstrated a recognition rate of 95.9% when tested on the FERET database [14]. Methods that use 3-D data for matching have to first resolve the alignment problem between the two face surfaces. Researchers developed an alignment model that takes multiple 2-D images of frontal and profile angles of the face and identifies key landmarks on the face [13]. This combination of images is then fitted to a universal model that is morphed to align with the landmarks extracted from the pair of images. Once a 3-D model is created, it is compared against the enrollment models using a number of different landmarks and a Euclidean distance between the landmarks is calculated to determine the degree of similarity. Algorithms based on 3-D shape data have used a variety of methods including surface contour matching, extended Gaussian images (EGI), comparison of shape volume, iterative closest point (ICP) matching, and PCA. A 3-D polygonal mesh representation is shown in Figure 4.6. Readers interested in the algorithmic details of these methods are encouraged to read [15].

4.4 Standards

The extensive use of face recognition in law enforcement, especially the FBI, has driven the development of face standards since the mid-1990s. The acquisition of face images does not depend on any specialized sensor; rather, it is significantly impacted by the quality of the camera, the environment, and user interaction. Instead of data representation, the various face standards have focused on eliminating variability in the face images.

The ANSI/NIST-ITL:2007 standard, originally published in 1986 for the exchange of fingerprint data between law enforcement agencies in the United States, created a section for standardized face data in 1997 [16]. The Type 10 record is designed specifically for facial mug shots, tattoos, or other identifying scars or marks. If the Type 10 record contains facial mug shots, optional information about pose, expression, the type of capture system, acquisition profile, compression type, and

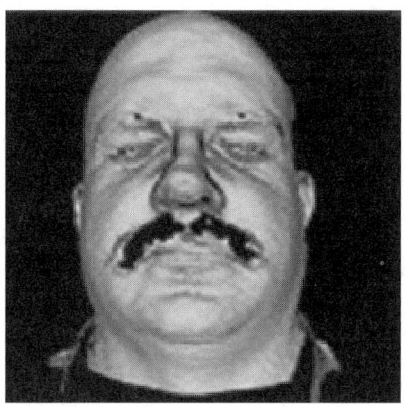 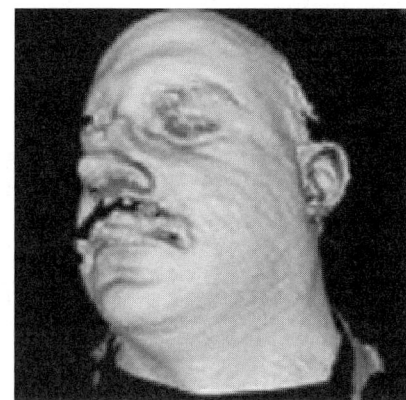

Figure 4.6 3-D rendering. (*Source:* E. Cabello, Rey Juan Carlos University. Reprinted with permission)

others can be encoded. The standard supports compression using baseline JPEG, JPEG2000, or PNG algorithms.

ISO/IEC 19794-5 standardizes data interchange elements capable of supporting human examiners as well as automated matching algorithms [6]. This standard has fields for physiological information such as gender, expression type, and eye color. A section of the standard is devoted to scene, photographic, and digital constraints, which were discussed in Section 4.2.2. The standard allows two different types of face images: full frontal and token. The difference between the two is the strictness of eye placement constraints enforced on each one. The full-frontal image is required to have a distance of at least 90 pixels between the eyes, whereas the token image is required to have the eyes at a specific coordinates that are related to the height and width of the image. By standardizing the eye location, token images reduce the effort of face detection and feature extraction and increase interoperability. An amendment was published in 2009, ISO/IEC 19794-5/Amd2:2009 [6], which standardizes data interchange for 3-D face data.

The ANSI INCITS 385-2004: Face Recognition Format for Data Interchange was published as a U.S. standard in 2004 [17]. This standard specifies the interchange of a face image with parameters for image dimensions, image resolution and focus, image color, and image characteristics. Face images are categorized into frontal and other types. Frontal images can be full frontal, which includes the face and shoulder line, or token, which includes only the face region.

The ISO/IEC TR 29794-5 technical report was published in 2010 to define and harmonize approaches for determining the covariates mentioned in Table 4.1, as well as normalize interpretation of quality scores from various vendors. Face image quality assessment is still an area that requires extensive research effort, especially in identifying universally accepted characteristics that can be measured consistently and are predictive of matching performance.

NIST published a document titled *Mobile ID Device Best Practice Recommendation* that provides guidelines for law enforcement agencies for capturing multiple biometric data, including face data, using mobile devices [18]. The best practices define image capture requirements for capture distance, image size, compression and type, capture device frame rate, capture sensor, and ambient conditions. The parameters for each capture requirement are mandated according to the intended use of the face image.

4.5 Evaluations

The development of automated face recognition received an impetus as independent, large-scale evaluations started in 1990s. This section will discuss some of the key evaluations that have led to advancement of 2-D and 3-D face recognition methods. A discussion of the face image databases collected for research purposes can be found in Chapter 13.

4.5.1 Facial Recognition Technology (FERET) Evaluations

NIST initiated the FERET program to assess the state of face recognition technologies and identify key areas of improvement for algorithm developers. The FERET program ran from 1993 to 1997, and three separate evaluations were conducted during that time. Due to the high concentration of academia and research groups working in this area, participation was limited to laboratory algorithms. The FERET evaluations had a significant impact on improving state of the art in face recognition as FRR (at fixed FMR = 0.001) reduced from 0.79 in 1993 to 0.54 in 1997. Along with its contribution to face recognition, the testing protocol used for FERET has provided the foundation for biometric system evaluation methodology. Readers interested in a complete analysis of the FERET results and evaluation methodology should read [19, 20]. The final dataset from the three FERET evaluations, which consists of over 14,000 images from more than 1,200 subjects, is available to researchers under license. Table 4.2 summarizes details of the FERET evaluations.

Table 4.2 FERET Summary

Evaluation	Number of Enrolled Individuals	Number of Algorithms	Description
FERET 1994	316	4	Fully automated algorithms
			Duplicate images taken on the same day with different poses and expressions
FERET 1995	817	3	Fully automated algorithms
			Duplicate images taken on the same day and on different days with different poses and expressions
FERET 1996	1,196	11	Included partially automated algorithms; landmark detection performed manually
			Duplicate images taken on the same day, on different days, and 1 year apart
			Introduced illumination variation along with pose and expression

Source: [19].

4.5.2 Face Recognition Vendor Test (FRVT)

By 2000 the face recognition industry was offering several commercially available systems. NIST started the FRVT program to assess the state of the art in commercially available systems. The FRVT program also ran as a series of three separate evaluations, starting in 2000, with each one introducing progressively more challenging real-world conditions. The FRVT 2000 was a technology evaluation that assessed the performance of commercial systems on various face image datasets and a scenario evaluation that assessed the usability of a complete face recognition system. The technology evaluation was performed to assess the impact of compression, resolution, distance between camera and subject, illumination, pose, and time lapse between the two images being compared. The results from the technology evaluation showed that progress had been made in the state-of-the-art systems in handling pose and illumination variation since the FERET evaluation, but the results were still not nearly acceptable for real-world deployments. Compression and resolution had an impact on performance, but are much easier to control and specify than the other performance covariates. The product usability evaluation assessed the final distance between the subject and the camera, the ability of the system to provide a result within 10 seconds, and the ability to acquire an image within 10 seconds. The product usability test highlighted the interoperability of cameras and the difference in enrollment and recognition environmental conditions as areas of future research [21].

The FRVT 2002 was conducted to evaluate the performance of face recognition systems on a large-scale dataset of images captured in real-world conditions. The test was divided into two parts: high computational intensity and medium computational intensity. In the high computational intensity test, all participants had to perform identification and verification on a dataset of around 121,000 full-frontal images collected in real-world conditions. The medium computational intensity test evaluated the capability of face recognition systems to compare still images under various illumination levels and pose conditions, as well as assess the impact of video data in face recognition. Ten systems were submitted for this test. The best verification rate was 90% at FAR of 1% for images captured indoors, whereas the best verification rate for images captured outdoors was 50% at FAR of 1%. Another observation of this test was that for every doubling of the database size, the top-ranked identification performance became worse by 2% to 3% [22]. FRVT 2002 established the expected performance for state-of-the-art systems in real-world conditions and identified outdoor images and size of enrollment database as key contributors to performance degradation.

The FRVT 2006 specifically evaluated high-resolution still 2-D face images, 3-D face images, and performance of still images taken in controlled and uncontrolled illumination. For recognition performance on high-resolution images, an FRR of close to 0.01 at a threshold FAR of 0.001 was observed for three of the participating systems. For low-resolution images an average FRR of approximately 0.05 at a threshold FAR of 0.001 was observed. The best performing system in 3-D face recognition generated an FRR of approximately 0.001 at a threshold FAR of 0.001. The performance evaluation of comparing high-resolution controlled illuminated images to uncontrolled illuminated images generated an FRR of approximately 0.15 at a threshold FAR of 0.001. FRVT 2006 also used human examiners to compare the performance of humans and automated techniques on the ability to recognize images captured under different illumination levels. The experiment found that automated systems can outperform humans at an FAR in the range of 0.05, and the complete results for all 22 participants are publicly available [23].

4.5.3 Face Recognition Grand Challenge (FRGC)

The FRGC was conducted in 2004 with the objective of advancing face recognition technology designed to support the missions of various U.S. government agencies. The starting point of the FRGC was the average performance of 80% true verification rate obtained in the FRVT 2002 by commercially available systems on controlled, indoor face images. The FRGC set the challenge of achieving a true verification rate of 98% at FAR of 0.1%. The FRGC was conducted as a series of six experiments with an overarching aim of examining new preprocessing techniques and developing methods for matching high-resolution images and 3-D face models to assess the best means of achieving the goal of 98% true verification rate. The results from FRGC showed an improvement in 2-D face recognition technology from FERET and FRVT 2000 and 2002 evaluations, and 3-D face recognition demonstrated promising results with room for improvement. The experiments with controlled 2-D face recognition yielded results of 99.9% verification and that of 3-D–3-D matching yielded results of 97%. Readers interested in detailed results of FRGC are encouraged to read [24].

4.5.4 Multiple Biometric Grand Challenge (MBGC) and Multiple Biometric Evaluation (MBE)

The Multiple Biometric Grand Challenge (MBGC) program was initiated by NIST in 2008 with the goal of addressing face and iris recognition problems that are more common in operational data. For face modality MBGC Version 1 concentrated on the recognition of low- to medium-resolution face images, the face recognition of near infrared illumination and high definition video from portals, and unconstrained face recognition from still and video sequences. MBGC Version 2 collected new datasets using the same challenge areas as MBGC Version 1. The MBGC program was a self-selected evaluation where participants were provided with biometric data and voluntarily provided self-reported similarity scores for the various datasets. The results from MBGC 1 and MBGC 2 are publicly available [25].

The Multiple Biometric Evaluation (MBE) program was initiated by NIST as an open evaluation of face and iris recognition systems based on the still images and portal video sequences. The MBE still face recognition supported four different test scenarios:

1. Verification without an enrollment database;
2. Verification with an enrollment database;
3. Identification with enrollment database containing up to 3 million records;
4. Pose calibration to assess the impact of face orientation.

At the time of this writing, only results from the MBE still face recognition test were available. The results showed that the most accurate face recognition algorithm had a 92% chance of identifying the unknown subject at rank 1 in a database of 1.6 million records, and the identification chance decreased as the size of the database increased. When human examiners are used in conjunction with the most accurate recognition algorithm, there was a 97% chance of identifying an individual when the top 50 ranked observations are returned against a database of 1.6 million records. Another interesting observation was that face recognition algorithms were more accurate on visa images compared to criminal mug shots, which can be attributed to a higher level of willingness for visa applicants to follow instructions and provide the best image possible [26].

4.5.5 Human-Machine Evaluation

Researchers have compared the performance of human face recognition with automated face recognition techniques [27]. Three different test scenarios based on exposure time were conducted in this experiment. The first scenario allowed examiners to view images for an unlimited time, the second scenario allowed examiners to view images for 2 minutes, and the third scenario allowed examiners to view images for 500 ms. The results showed that human examiners were no more accurate when allowed to view face images for more than 2 minutes, but an exposure time of 500 ms had a significant degradation in performance compared to other scenarios. A total of seven algorithms were evaluated in this study; out of these algorithms, three showed better performance than human examiners. The results demonstrated that human examiners are comparable with the current state-of-the-art face recognition systems. Another research conducted on 338 full-frontal images of 131 individuals showed that 29.2% humans performed better than a commercial face recognition system from 2006, while 37.5% humans performed worse [28].

4.5.6 Other Evaluations

The Face Verification Competition was conducted in 2004 and was open for public participation [29]. The competition was conducted in two separate phases. In the first phase a dataset consisting of normalized and localized face images was used. The second phase dataset consisted of unconstrained nonlocalized and nonnormalized images. A performance metric called a *weighted error rate* (WER) was calculated as a function of FAR and FRR at a specific threshold and three different weights for FAR and FRR. The results showed that preprocessed face images improved performance significantly, which supported the need for improved face localization and normalization in face recognition systems. Table 4.3 summarizes the results for the six systems that participated in the evaluation.

An evaluation of commercial systems on the face recognition of twins generated an EER of approximately 5% when images were captured under controlled lighting and neutral expression [30]. Images were collected from 126 pairs of twins for this evaluation. When uncontrolled images of 24 pairs of twins after a year were captured, the EER of the best system was observed to be approximately 17%. This research has brought focus onto an area that is relatively unexplored.

The Mobile Biometry (MOBIO) database was collected as part of a multiple biometric evaluation on video face and speaker recognition using a mobile device. The performance of the face recognition systems was calculated using a metric

Table 4.3 Summary of Face Verification Competition 2004

System	Phase 1 Average WER %	Phase 2 Average WER %
1	3.53	3.78
2	2.7	5.5
3	3.5	7.88
4	1.95	4.4
5	2.99	3.93
6	3.04	4.18

Source: [29].

called a *half total error rate* (HTER), which is an average of the FAR and FRR at a specific threshold. The MOBIO face recognition results were categorized based on gender and are summarized in Table 4.4.

There are several prominent datasets that have been collected and released for researching pose, illumination, expression, aging, and other demographics. A more detailed discussion of these datasets is given in Chapter 13.

4.6 Applications and Trends

The majority of automated face recognition deployments are still in law enforcement with a relatively small proportion in border control and consumer applications. This section will discuss some of these successful systems.

4.6.1 Law Enforcement

Law enforcement can take the most advantage of automated face recognition due to the existence of a large volume of face images. The Pinellas County Sheriff's Office in Florida has deployed automated face recognition at the booking center and the courthouse. Individuals brought into custody are screened against the existing face database for prior records. If a record does not exist, a new one is created that is

Table 4.4 MOBIO Face Recognition Results (HTER %)

Algorithm	Male	Female	Average
1	25.45	24.39	24.92
2	16.92	17.85	17.38
3	25.43	20.83	23.13
4	31.36	29.08	30.22
5	9.75	12.07	10.91
6	10.30	14.95	12.62
7	29.80	23.89	26.85
8	20.50	27.26	23.88
9	21.86	23.84	22.85

Source: [31].

then used for checks during the formal booking process and release process. Such systems have been extended to several counties across the United States, although data sharing is still an operational and policy challenge.

4.6.2 E-Passport

In 2002 the International Civil Aviation Organization (ICAO) passed a resolution that endorsed the use of automated face recognition as part of the passport authentication process. E-passports that are biometric-enabled require a digital image of the passport owner to be embedded per the ICAO specifications. Countries that have the capability of automated face verification at the point of entry can use this face image to ensure that the true owner is in possession of the passport. E-passports have been issued by several countries since 2006, although automated face verification is not yet enforced.

4.6.3 U.S. Department of State Visa Application System

As part of the Enhanced Border Security and Visa Entry Reform Act of 2002, the U.S. Congress mandated the use of biometrics in U.S. visas. In 2004 the U.S. Department of State started integrating automated facial recognition into their visa application process to screen for duplicates, aliases, and individuals on a watch list prior to the issuance of a visa. This is also called the BioVisa program. The system enrolled more than 30 million legacy visa applicants and has been enrolling every new visa applicant since.

4.6.4 Surveillance Applications

The use of automated face recognition is a natural extension of video surveillance, as it can be deployed covertly. Assisting human examiners with automated face recognition can increase the efficiency of the surveillance program. For instance, all individuals entering a certain area can be screened against a watch list with the aid of automated face recognition, and any probable hits can be passed onto human examiners for final determination. Casinos have a long history of using automated facial recognition for the surveillance of both employees and patrons.

4.6.5 Logical Access

Logical access to computing and network resources can be controlled using face recognition due to the increasing integration of cameras in handheld devices and computers. Several computer vendors have started providing face recognition in laptops as a replacement for passwords and this trend will likely increase.

4.6.6 Airport Applications

There are various airports around the world using face recognition for identifying employees, airline staff, and frequent fliers. The SmartGate system is used by the Australian Customs and Border Protection Service and the New Zealand Customs Service for self-processing eligible travelers through the immigration checks. It was first deployed at Sydney International Airport, Australia, in 2002 and used for the automated recognition of frequent fliers and Quantas airline staff, and it has since expanded its operations. The Lyon-Saint Exupéry Airport in France has been using 3-D face recognition since 2006 for the authentication of pilots and employees who have access to the airport tarmac. The Sheremetyevo International Airport in Moscow, Russia, has been using 3-D face recognition since 2010 to authenticate airport staff. These deployments are indicative of the potential of automated face recognition, especially in airport security and border control.

4.6.7 Personal Applications

The growth of online social networks such as Facebook has spawned an innovative use for face recognition in the online arena. Applications such as Facebook use facial recognition software to automatically identify friends in images and use this information to suggest new people who can be added to the users' network. With an increasing amount of information and a significant amount of images being put online, face recognition will undergo a proliferation to automate processes that require correlating multiple data sources.

4.7 Design and Deployment Considerations

Face recognition systems offer a level of flexibility and convenience that is currently unmatched by any other biometric technology. Face acquisition is a process that humans understand quite well and its familiarity reduces the barrier for adoption. However, in order to achieve an acceptable performance, constraints are required for several aspects of the system. This section will discuss factors that should be taken into account for designing and deploying a face recognition system.

4.7.1 Pose and Expression

Research has shown that pose and expression have a significant impact on the performance of 2-D face recognition systems. Commercial systems are capable of handling minor pose variations and changes in expression, but performance degradation is inevitable in an uncontrolled scenario. Appendix A of ISO/IEC 19794-5 includes a list of best practices for image capture that recommends controlling the pitch, yaw, and roll of a face pose within ±5°, as well as a neutral face expression with open eyes [6]. These recommendations are easier to enforce in a supervised system. Unsupervised systems require a user interface that can guide the users in an intuitive manner. A generic face outline on the user interface that aids the positioning of the user with respect to the camera is an effective meas of reducing pose challenges.

4.7.2 Physiological Factors

A clear view of the face is necessary in order to locate landmarks for the alignment and extraction of face features. Any occlusion due to artifacts such as sunglasses, hats, or clothing will interfere with face detection and needs to be addressed as part of the acquisition process. Physiological changes due to short-term aging do not have a significant impact on face recognition, but long-term aging studies have yet to be conducted to determine their effect. A template update or adaption method that is performed at regular intervals can reduce aging effects on face recognition. A drastic change in the facial structure is not possible without undergoing a complex surgical procedure, but for 2-D face recognition facial hair such as a beard or a mustache can have an impact on landmark detection. Newer systems are robust at handling facial hair, but a deployment should examine this as part of a pilot.

4.7.3 Environmental Factors

A change in illumination affects the reflectance properties of the skin and the resulting intensity information captured by the camera. For 2-D face recognition, which depends heavily on pixel intensities for processing face images, illumination changes can significantly impact performance. Hot spots and uneven distribution of light are examples of illumination effects that interfere with feature detection and extraction. The amount of noise and clutter in the image background is an issue for the segmentation of the face. The presence of multiple faces in the background can confuse the system about the target face, although newer systems are capable of segmenting multiple faces from an image.

4.7.4 Image Compression and Interoperability

Data compression is an important consideration from a transmission and storage perspective. Networked systems rely on quick transmission of data to reduce the total processing time. Handheld devices have limited storage capacity. A key issue is to determine the impact of image format and compression levels on performance and quantifying the degradation in a predictable manner. There are several different image formats and compression algorithms such as PNG, JPEG2000, PNG, and others. Analysis in ISO/IEC 19794-5 reported that performance degrades quickly for face images compressed to sizes less than 10 KB and JPEG2000 compression is more effective than JPEG compression [6]. The analysis also showed that color images tend to perform better than grayscale images. It should be noted that these results are a function of the face recognition system used in the analysis and these results should not be generalized across all face recognition systems.

Although data interchange formats have been specified in various standards, comparing images compressed using different algorithms and compression levels raises a real-world interoperability issue that has received scant attention. A system might require using the same compression level and algorithm for all its enrolled face images, which could lead to interoperability issues when face data needs to be exchanged with other agencies. In such cases the impact of image compression on matching performance and methods to compensate for performance degradation become very important.

Readers interested in the impact of image compression on face recognition are encouraged to read [32].

4.7.5 Subject Camera Distance and Motion

The distance between the camera and subject has an impact on the resolution of the image and ultimately on the performance of the system. Based on empirical results that analyzed the impact of intereye distance on verification performance, ISO/IEC 19794-5 has mandated a distance of at least 180 pixels across the width of the head and 90 pixels between the eyes as a conformance clause for its full-frontal image type [6]. Relative motion between the subject and the camera can introduce blurring effects in the resulting image and degrade performance. Providing feedback to users about optimal positioning through either audio or visual methods is extremely important for improving the performance of real-world systems.

4.7.6 Spoofing Attacks

The most commonly executed spoofing attack is to use a photograph of a subject during the acquisition process. In 2009 researchers demonstrated a successful spoofing attack on face recognition systems that were available with three commercially available laptops [33]. To address these spoofing issues, at the time of this writing, an open competition called the Competition on Counter Measures to 2D Facial Spoofing Attacks is being conducted. Results were expected to be announced at the International Joint Conference on Biometrics at the end of 2011. The objective of this competition is to examine the performance of spoof detection algorithms for face images.

From a deployment perspective a challenge response interaction should be a part of the acquisition process to ensure that a real person is providing the face image. An example of a challenge response is asking the user to move his or her head from side to side and determining user liveness.

4.8 Summary

Automated face recognition can be used in a diverse range of applications, and although studies over the last decade have exposed a gap between expectations and operational performance of face recognition systems, it will keep evolving with advances in technology. Face recognition beyond the visible spectrum has received little attention; multispectral face imaging techniques will start receiving increased attention to fulfill the needs of highly specialized applications. Face recognition at a distance is an area of interest, especially in military applications. Face recognition using mobile devices is expected to grow as these devices become cheaper and more powerful. Law enforcement uses a combination of human experts and automated recognition techniques. The ability to compare a hand-drawn sketch against a database of images using automated techniques is of great interest, and ongoing research is examining this particular capability. Building automated recognition algorithms modeled on the hierarchical methods used by humans also is a fertile research area. The 3-D face recognition has now entered the home environment with the release of Xbox Kinect. This device also provides a range of images to independent developers that will open up research possibilities in 3-D face recognition.

4.8 Summary

Face recognition has advanced as different aspects of the problem have been analyzed since the early 1990s. Interest from the law enforcement and intelligence communities has largely driven progress of this technology, with a relatively smaller proportion of deployments in consumer-facing applications. Although the performance of automated techniques has not been able to keep up with the expectations of users, new methods using 3-D methods are closing the gap. Face recognition offers certain advantages that other biometric technologies cannot offer such as intuitive interaction, familiarity, and nonphysical contact, and these will keep face recognition relevant in the near future. Along with border control applications, face recognition will likely see an increase in specialized applications for military and surveillance.

References

[1] Bledsoe, W. W., "Some Results on Multicategory Patten Recognition," *Journal of the Association for Computing Machinery*, Vol. 13, 1966, pp. 304–316.

[2] Bledsoe W. W., and H. Chan, *A Man-Machine Facial Recognition System—Some Preliminary Results*, Technical Report, Panoramic Research, Inc., Palo Alto, CA, 1965.

[3] Sirovich, L., and M. Kirby, "A Low-Dimensional Procedure for the Characterization of Human Faces," *Journal of Optical Society of America*, Vol. 4, 1987, pp. 519–524.

[4] Turk, M. A., and A. P. Pentland, "Face Recognition Using Eigenfaces," *Proceedings of the IEEE*, Vol. 76, 1991, pp. 586–591.

[5] Blanz, V., and T. Vetter, "A Morphable Model for the Synthesis of 3D Faces," *SIGGRAPH 99*, 1999, pp. 187–194.

[6] ISO/IEC, *ISO/IEC 19794-5:2005—Biometric Data Interchange Formats—Part 5: Face Image Data*, Geneva, Switzerland, 2005.

[7] ISO/IEC, *ISO/IEC TR 29794-5:2010—Biometric Sample Quality—Part 5: Face Image Data*, Geneva, Switzerland, 2010.

[8] Wiskott, L., et al., "A Morphable Model for the Synthesis of 3D Faces," *IEEE Transactions on Pattern Analysis and Machine Intelligence*, Vol. 19, 1997, pp. 775–779.

[9] Belhumeur, P., J. Hespanha, and D. Kriegman, "Eigenfaces vs Fisherfaces: Recognition Using Class Specific Linear Projection," *IEEE Transactions on Pattern Analysis and Machine Intelligence*, Vol. 19, 1997, pp. 711–720.

[10] Bartlett, M., and T. Sejnowski, "Independent Components of Face Images: A Representation for Face Recognition," *Proc. 4th Annual J. Symp. Neural Computation*, 1997.

[11] Gao, Y., and M. K. H. Leung, "Face Recognition Using Line Edge Map," *IEEE Transactions on Pattern Analysis and Machine Intelligence*, Vol. 24, 2002, pp. 764–779.

[12] Zhao, W., et al., "Face Recognition," *ACM Computing Surveys*, Vol. 35, December 2003, pp. 399–458.

[13] Ansari, A. -N., M. Abdel-Mottaleb, and M. H. Mahoor, "A Multimodal Approach for 3D Face Modeling and Recognition Using 3D Deformable Facial Mask," *Machine Vision and Applications*, Vol. 20, January 2008, pp. 189–203.

[14] Volken, B., and T. Vetter, "Face Recognition Based on Fitting a 3D Morphable Model," *IEEE PAMI*, Vol. 25, No. 9, September 2003.

[15] Bowyer, K., K. Chang, and P. Flynn, "A Survey of Approaches and Challenges in 3D and Multi-Modal 3D+2D Face Recognition," *Computer Vision and Image Understanding*, Vol. 101, January 2006, pp. 1–15.

[16] NIST, *American National Standard for Information Systems- Data Format for the Interchange of Fingerprint Facial, & Other Biometric Information—Part 1*, 2007.

[17] M1, *ANSI INCITS 385-2004-Information Technology—Face Recognition Format for Data Interchange*, 2004.

[18] Orandi, S., and R. M. McCabe, *Mobile ID Device Best Practice Recommendation Version 1*, Gaithersburg, MD: National Institute of Standards and Technology, 2009.

[19] Phillips, P., P. Rauss, and S. Der, *FERET (Face Recognition Technology) Recognition Algorithm Development and Test Report*, 1996, p. 73.

[20] Phillips, P. J., et al., "An Introduction to Evaluating Biometric Systems," *Computer*, 2000, pp. 56–63.

[21] Blackburn, D. M., "The Design and Implementation of the Facial Recognition Vendor Test 2000 Evaluation Methodology," M.S. thesis, Virginia Polytechnic Institute and State University, 2001.

[22] Phillips, P., et al., *Face Recognition Vendor Test 2002*, NIST and DD Report, 2003, p. 278.

[23] Phillips, P. J., et al., "FRVT 2006 and ICE 2006 Large-Scale Experimental Results," *IEEE Transactions on Pattern Analysis and Machine Intelligence*, Vol. 32, May 2010, pp. 831–846.

[24] Phillips, W. W. P. J., et al., "Preliminary Face Recognition Grand Challenge Results," *7th International Conference on Automatic Face and Gesture Recognition*, 2006, pp. 15–24.

[25] NIST, "MBGC Presentations & Publications," *NIST*, 2010.

[26] Grother, P. J., G. W. Quinn, and P. J. Phillips, *Report on the Evaluation of 2D Still-Image Face Recognition Algorithms*, Gaithersburg, MD, 2011.

[27] O'Toole, A. J., *Human Versus Machine Performance*, NIST Technical Report, 2006.

[28] Adler, A., and M. E. Schuckers, "Comparing Human and Automatic Face Recognition Performance," *IEEE Transactions on Systems, Man, and Cybernetics*, Vol. 37, 2009, pp. 1248–1255.

[29] Kittler, J., et al., "Face Authentication Competition on the BANCA Database," *Proceedings of the International Conference on Biometric Authentication (ICBA)*, Hong Kong, 2004, pp. 15–17.

[30] Strickland, E., "Can Biometrics ID an Identical Twin?" *IEEE Spectrum*, 2011.

[31] Marcel, S., et al., "On the Results of the First Mobile Biometry (MOBIO) Face and Speaker Verification Evaluation," *Lecture Notes on Computer Science (ICPR 2010 Contest Series)*, 2010.

[32] Delac, K., S. Grgic, and M. Grgic, "Image Compression in Face Recognition: A Literature Survey," in *Recent Advances in Face Recognition*, K. Delac, M. Grgic, and M. S. Bartlett, (eds.), InTech, 2008, pp. 1–14.

[33] HSNW, "Researchers Spoof, Bypass Face-Recognition Authentication Systems," *HSNW*, 2009.

CHAPTER 5

Iris Recognition

Iris recognition uses the texture pattern on the surface of the iris for recognition. The iris is the colored ring surrounding the pupil, which is the dark circle at the center of the eye that controls the amount of light entering the eye. Iris recognition has seen tremendous growth in the last decade, both in terms of research and commercialization, and it has been deployed successfully in large-scale applications around the world. Iris recognition has proven to be a robust biometric technology and its use will only increase as large-scale identification programs, such as national identification and voter registration, increasingly use biometrics. This chapter will discuss the workings of each iris subsystem and their operational challenges and provide a framework for understanding various deployment scenarios.

5.1 Anatomy of Iris

The visible part of the human eye is comprised of the sclera, iris, and pupil. The outer white part of the eye is the sclera, which is comprised of connective tissue and blood vessels. The inner dark part of the eye is the pupil and it is responsible for controlling the amount of light entering the eye. The colored middle part of the eye is the iris and it is a muscle tissue and the only internal human organ that can be seen from outside the body. The anterior region of the iris, called the *stroma*, is divided into two main zones: the pupillary zone and the ciliary zone. The pupillary zone is the part of the iris that forms a boundary with the pupil. The rest of the iris that extends out to the sclera is the ciliary zone. The area between the pupillary zone and the ciliary zone is called the *collarette*. The muscle fibers in the stroma have openings in them for biological functions that result in structures called *crypts*. These, along with furrows found in the pupillary and ciliary zones, are features used for iris recognition. These features are shown in Figure 5.1. The formation of the iris is heavily influenced by the gestation conditions and studies have confirmed that they have a high degree of discriminating power between individuals. Experiments have shown that iris patterns are different between eyes of the same individual as well as between twins [1]. The eye is an extremely well-protected organ that reduces the chances of injuries. These factors make the iris an ideal physical trait for biometric recognition.

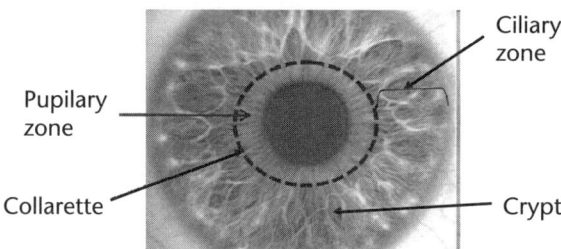

Figure 5.1 Iris with regions identified. (*Source:* J. Daugman, University of Cambridge. Reprinted iwth permission.)

5.2 History

Alphonse Bertillon suggested the use of the iris for differentiating between individuals in 1886 while conducting his experiments on anthropometrics, and in 1936 ophthalmologist Frank Burch came to a similar conclusion in the course of treating his patients [2]. The first patent on iris recognition was awarded to two ophthalmologists in 1987, Leonard Flom and Aran Safir, where they described the theoretical processes for an operational iris recognition system [3]. In 1992 the Los Alamos National Laboratory published a report outlining the feasibility of iris images for recognition purposes [4]. In this report iris images were collected from 650 individuals over a period of 15 months and it was concluded that iris images remained unchanged for the duration of the study and contained information that could differentiate between individuals, but these results were not supported with any experimental results. In 1994 John Daugman was awarded a patent in which he described an operational iris recognition system in detail [5]. His patent described the processes of each subsystem and his approach was validated with experimental results that used iris images acquired from live subjects. Daugman's work is seminal in the area of iris recognition and terms such as IrisCode that were used in his early publications are now standard. Although other techniques for iris recognition have been published since then, Daugman's approach forms the basis for a majority of the commercially available iris recognition systems. A company called IriScan, which has since undergone several ownership and name changes, created the first commercial iris recognition system based on the patents of Flom-Safir and Daugman in 1995. The Flom-Safir patent expired in 2005, and Daugman's patent was scheduled to expire in 2011, which industry analysts expect will increase the number of iris recognition vendors.

5.3 Iris Image Acquisition

Most commercial iris recognition systems use near infrared (NIR) illumination for capturing iris images (see Figure 5.2). NIR wavelength falls in the 700–900-nanometer (nm) range and has several operational advantages over visible light:

1. NIR light is not visible to the human eye and does not cause any discomfort when directed at the human eye.

5.3 Iris Image Acquisition

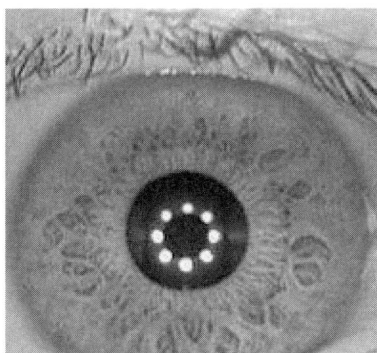

Figure 5.2 Infrared iris image. (*Source:* CASIA-Iris-Interval Database, Chinese Academy of Sciences. Reprinted with permission.)

2. NIR light reflection off the cornea is less disruptive than visible light during capture.
3. Melanin pigment in the iris provides color to it. Melanin absorbs most of the visible light, but reflects most of the NIR light. Illuminating the eye with NIR light isolates the texture information. Although, from an information theory perspective, a color iris image has more information than an NIR illuminated image, in practice NIR illumination can capture texture information from dark irises more effectively than visible light.
4. Controlling NIR illumination is much easier than ambient light and thus provides greater control over the capture process.

In most iris recognition systems light emitting diodes (LED) are used for infrared-illumination. Some systems use multiple sets of LEDs that provide NIR illumination of different wavelengths. The boundary between sclera and iris is easy to determine in visible light due to the difference in contrast, but NIR-illuminated images do not provide a high level of contrast. NIR illumination between 750 and 800 nm provides a better boundary differentiation between sclera and iris, whereas 850–900 nm provides better iris texture information. Some systems capture images at both wavelengths and then use the appropriately illuminated image for specific image extraction operations.

The earliest commercial iris systems typically required users to position themselves within 4–12 inches of the camera and could capture only one iris image at a time. Another limitation of these systems was its physical size. Iris cameras capable of capturing images beyond approximately 1 foot were suitable only for physical access control and smaller-sized iris cameras required that they be connected to a fully functional computer. Over the years iris imaging technology has steadily improved, which has resulted in mobile iris systems and standoff iris systems. Mobile iris systems have compressed the imaging system to a form factor that is small enough to fit into small-scale devices such as today's smartphones. Such systems require users to be within 4–8 inches of the device. By contrast, standoff iris systems capture iris images from beyond 1 meter. As one can imagine, iris acquisition is the biggest challenge for standoff iris systems and it is discussed in Section 5.6. Readers interested in the technical aspects of iris acquisition are encouraged to read Matey and Kennell [6].

Since Daugman's work is the most influential in the field of iris recognition, the feature extraction and feature matching sections will primarily discuss his methodology. Research references to other methodologies are given for interested readers.

5.4 Feature Extraction

The first step in feature extraction is iris segmentation, which entails isolating the iris from the entire image. Iris segmentation is an easy process to understand; the goal is the accurate boundary detection of the pupil, sclera, and eyelids. Figure 5.3 illustrates the feature extraction process. However, automating this process is one of the most challenging problems in iris recognition as an incorrectly segmented iris increases the probability of a matching error. There are several conditions that lead to segmentation errors: unless the eye is opened up extremely wide, the iris is occluded by eyelids and eyelashes; inconsistent illumination can reduce the contrast between the pupil, iris, and sclera or an extremely dark iris might not be differentiated from the pupil. Figure 5.4 shows a few examples of nonideal iris images. Commercial vendors heavily guard segmentation as a critical component of their intellectual property. Earlier segmentation algorithms used approximation models

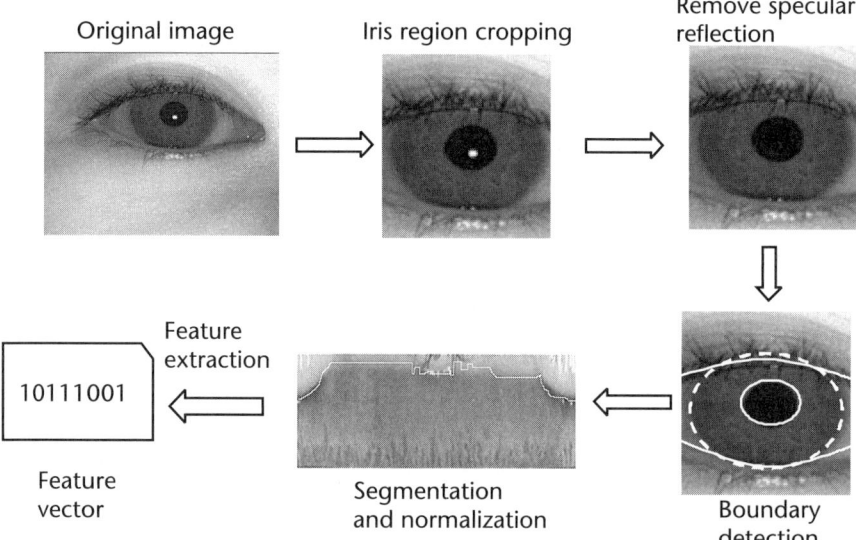

Figure 5.3 General iris extraction process.

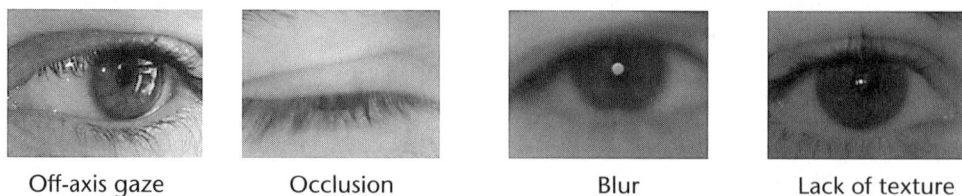

Figure 5.4 Examples of nonideal iris images. (*Source:* H. Proenca, UBIRIS Database, University of Beira Interior, Portugal. Reprinted with permission.)

to fit the boundaries between the pupil, iris, and sclera as circles, although recent research has shown a better fit with noncircular models. This assumption of symmetry laid the foundation for most of the earliest work conducted in iris segmentation. In earlier algorithms the circles were described using three parameters: x-coordinate of center, y-coordinate of the center, and radius. In his early works Daugman described a segmentation algorithm, also called Iris2pi, which used an integro-differential operator over the entire parameter space to detect the boundaries. The integro-differential operator and high-level details were described in Daugman's 2004 paper, but the internal details are not publicly available. In 2007 Daugman explored new techniques for modeling the pupil and sclera boundaries as ellipses, as it is thought to be a more accurate representation [7]. Another set of iris segmentation algorithms stemmed from the original work conducted by Wildes [8]. This approach uses edge detection techniques in combination with Hough transform to detect the best fit for pupil and sclera boundaries. The goal of the Hough transform is to approximate a circle based on a set of edge points. The body of work in this area is quite vast, and readers are encouraged to read Bowyer et al. [9].

Once the iris image is segmented, a normalized template is generated. Iris images captured over multiple acquisitions will not be the same size for a variety of reasons: the distance between the iris and camera will never be exactly the same and the illumination intensity directly impacts the size of the pupil, and indirectly impacts the iris. Daugman's approach converts the image into polar coordinates so that every point can be represented using an angular parameter between 0 and 360, and a radial parameter. The circular iris image is unwrapped into a rectangular image where the vertical axis of the image represents the angular boundary and the horizontal axis of the image represents the radial boundary, and each point can be described using polar coordinates. The normalized image is then divided into a grid of 128×8 blocks, which results in 128 vertical divisions and 8 horizontal divisions. Each block is convolved with a 2-D Gabor filter to extract texture information, and the filter's phase response is measured. Phase information is used because it is more discriminating and invariant than amplitude, which depends on contrast information in the image. The phase angle is a complex number (comprising a real component and an imaginary component) and it is quantized to 2 bits to facilitate matching. If the real component of the coefficient is positive, it is represented as 1 or else as 0. If the imaginary component of the coefficient is positive, it is represented as 1 or else as 0. The phase angle is thus represented using 2 bits, which results in the eventual iris template consisting of: $(128 \times 8) \times 2$ bits = 2,048 bits = 256 bytes. This 256-byte template is also called an IrisCode. A corresponding mask array is also computed along with the iris template that indicates if the texture information in that block met the quality criteria. The mask array is used to skip noisy blocks during the matching process.

5.4.1 Quality Assessment

The development of an effective iris image quality metric has begun receiving increased attention in the last 5 years. There is no absolute consensus on parameters required for iris image quality assessment, but research has shown that focus, motion blur, occlusion level, radius ratio of pupil and iris, and sight orientation play an important role. Table 5.1 lists the covariates that research has identified as having an impact on iris recognition.

The year 2009 marked the initiation of several standards-related initiatives in the iris image quality. The ISO/IEC 29794-6 project, currently in development, was initiated in 2009 to define methods used to characterize and assess iris image quality. NIST also initiated the IREX-II project with the specific aim of defining and measuring iris image quality indicating its importance to the standardization community, especially for large-scale deployments [10]. Image quality assessment is directly tied to improving feedback for operators and users, as well as matching performance. Current commercial systems come equipped with proprietary image quality assessment modules that are based on some of the factors discussed in Table 5.1.

5.5 Iris Feature Matching

Daugman's approach computes a metric called the *normalized Hamming distance* (HD) for the comparison of two IrisCodes. For each corresponding bit in the two IrisCodes, their corresponding mask bits are checked for quality. If both mask bits indicate that the IrisCode bits are usable, they are compared. This process is performed over the entire IrisCode, and the number of bits that disagree is computed. The final score is computed from the resulting fraction of bits that disagree, after the number of unusable bits have been discarded. This score is compared to a system threshold that results in a decision. In most Daugman-based systems, this threshold tends to be around 0.33, and a successful match is declared for a similarity score of less than 0.33. For example, a score of 0.28 indicates that 28% of the bits disagree between the two IrisCodes. Large-scale studies and statistical analysis has shown that this threshold is acceptable for operational systems.

Daugman's matching algorithm has several benefits. It is extremely quick since the comparison is a bitwise operation. The rotation of the iris image needs to be only accounted for during the matching operation. The bit representation of one IrisCode needs to be shifted relative to the other IrisCode and then compared to compensate for any rotational variation.

Table 5.1 Image and User Covariates Affecting Iris Recognition

Acquisition Covariates	*Subject Covariates*
Image defocus	Eye color
Dedicated illumination	Eye wear
Dynamic range	Intrinsic anatomical features: contrast between boundaries, applicability of circular iris model
Motion blur	Face orientation
Occlusion	Face rotation
Optical resolution	Sight orientation
Pixel sampling	Pupil size

Source: [11].

5.6 Standards

Apart from the ongoing project on iris image quality standard, there is an internationally accepted standard for exchanging iris image data. The ISO/IEC 19794-6:2005 specifies two different formats for exchanging iris data: a rectilinear image represented in Cartesian coordinates (x, y) that can have either uncompressed, raw, or compressed intensity values, and a polar image represented in polar coordinate (r, θ) [12]. Although the polar format is much smaller than the rectilinear format as it only contains iris information, this format was shown to be sensitive to the localization method used for segmentation [13]. In 2008 the ISO/IEC SC 37 committee decided to remove the polar coordinate format from all future revisions of the standard, and NIST published the IREX I report, which contained an evaluation of three new proposals of compact interoperable iris data formats (see Section 5.7). The latest update to ISO/IEC 19794-6 will take these recommendations into account, and it remains to be seen which compact formats will be accepted. The U.S. version of the iris standard ANSI/INCITS 379:2004 was replaced by the ISO/IEC 19794-6 in 2008.

The ISO/IEC 29794-6 project, which was still in the drafting stages at the time of this writing, was initiated in 2009 to define methods used to characterize and assess iris image quality. This standard will specify covariates that influence iris image quality as well as provide context for quality scores to improve interoperability.

The ANSI/NIST-ITL:2007 standard is used for data exchange between law enforcement agencies [14]. It uses a generic iris image data with optional fields that can be populated using information from ISO/IEC 19794-6:2005. The standard allows use of baseline JPEG or JPEG2000 algorithm for compressing iris images at a maximum compression ratio of 6:1.

NIST published a document titled *Mobile ID Device Best Practice Recommendation* that provides guidelines for law enforcement agencies for capturing multiple biometric data, including iris images, using mobile devices [15]. The best practices define acquisition requirements for iris diameter, exposure time, feedback for image quality, capture volume, capture distance, image wavelength range, scan type, image margins around the iris border, image sampling frame rate, allowable maximum average irradiance, sensor signal-to-noise ratio, pixel depth, and image compression. The parameter for each capture requirement is mandated according to the intended use of the iris image.

5.7 Iris Capture at a Distance

Iris recognition systems have started stretching their capture capability to beyond 1 meter and farther to try and recognize individuals at a distance. Such systems are generally called *standoff iris recognition systems* and some are capable of capturing iris images from individuals in motion. The main challenge in standoff iris systems arises from the acquisition phase. Iris capture at a distance is mainly affected by the following:

1. *Physical distance:* Infrared light dissipates over long distances and it needs to have a higher level of energy to travel such distances. Infrared

illumination with a high level of energy can be harmful to the eye and user safety is a critical consideration.

2. *Capture volume:* This is the invisible space that defines the area in which a user should stand for his or her iris to be captured.
3. *Relative motion:* Iris systems are capable of capturing an iris image from a moving subject. These systems have to compensate for motion blur and off-axis gaze in addition to having an extremely quick capture mechanism.
4. *Environment:* The further the distance between the camera and the subject, the higher the chance of environmental conditions impacting the capture process. It is difficult to control ambient lighting over a large distance.
5. *Safety:* High-intensity infrared illumination can be harmful to the human eye and limits the ability to capture clear images at a distance.

Standoff iris systems are gaining traction in surveillance applications and border control systems. These systems also provide the ability of capturing face images and using multimodal recognition to enhance performance. Readers who are interested in the technical details of standoff iris recognition systems are encouraged to read [6].

5.8 Evaluations

Performance evaluation of iris recognition has lagged behind fingerprint and face recognition, but a number of large-scale evaluations have examined a variety of performance influencing factors such as the interoperability of cameras and data formats, human factors, and the general state of the art. This section will discuss a small sample of these performance evaluation studies. Please note that the iris recognition industry has undergone several cycles of expansion and consolidation, and some of the companies mentioned in this section may have changed their names or might not be in existence anymore.

5.8.1 Independent Testing of Iris Recognition Technology (ITIRT)

The ITIRT, conducted by the International Biometric Group (IBG) in 2005, was among the first large-scale independent testing of iris systems. The objectives of the ITIRT was to examine acquisition and matching performance metrics of iris databases collected using three different iris cameras and analyze the level of effort for image acquisition [16]. This study used three different iris cameras: LG IrisAccess 3000, OKI IRISPASS-WG, and Panasonic BM-ET330 and template generation and matching platform provided by Iridian (now owned by L-1 Identity Solutions, Inc.). IBG reported acquisition FTA, FTE, FNMR, FMR, acquisition time, and interoperability performance rates. Table 5.2 provides a summary of the acquisition rates, and Table 5.3 provides a summary of the native and interoperability attempt level error rates.

This study had a number of interesting conclusions:

5.8 Evaluations

Table 5.2 ITIRT Performance Metrics for the Acquisition of at Least One Iris

System	FTA%	FTE %	Enrollment Time (sec)
LG	1.77	8.5	69.79
OKI	0.91	7.79	118.13
Panasonic	1.09	9.63	63.78

Source: [16].

Table 5.3 ITIRT Native and Interoperability Transaction Error Rates at a Threshold of 0.330 HD

System		LG	OKI	Panasonic
LG	FNMR %	1.570000	2.454000	2.301000
	FMR %	0.008700	0.001610	0.001990
OKI	FNMR %	3.240000	1.038000	2.297000
	FMR %	0.000900	0.000350	0.000620
Panasonic	FNMR %	2.676000	2.641000	0.583000
	FMR %	0.001540	0.000830	0.001290

Source: [16].

- The comparison of iris images from different cameras indicated that native performance was better than interoperable performance.
- The iris images from different eyes of the same person showed a higher level of similarity than the iris images of different persons, indicating a higher correlation in the iris images of different eyes of the same person. However, it should be noted that the similarity was not close enough to declare a false match.
- The compression of the iris images had a degradation impact on the similarity scores of 99% of the comparisons.

The ITIRT shed light on a number of challenges in iris recognition and significantly influenced some of the studies that followed.

5.8.2 Iris Challenge Evaluation (ICE)

The ICE was a two-phase study conducted by NIST in collaboration with several universities and government agencies to examine the current state of iris recognition and identify areas of future need. ICE 2005, the first phase of the study, used data collected from 132 subjects at the University of Notre Dame on an LG 2200 iris camera. The main purpose of ICE 2005 was to evaluate right-eye and left-eye independence in terms of iris quality and recognition performance. Two different experiments were conducted for this purpose: right iris compared to right iris and left iris compared to left iris. ICE 2005 was conducted as a self-selected evaluation in which nine participants agreed to process this data using their own iris feature extraction and matching algorithm and provide matching lists back to NIST. NIST generated ROC curves for all nine participants and observed that performance was quite varied between the best and worst performing algorithms. The top performers reported a true verification rate of 0.995 at an FAR of 0.001 for the right eye, and a true verification rate of 0.990 to 0.995 at an FAR of 0.001 for the left eye. A majority of the systems performed better on the right iris images than on the left iris images. The results showed a correlation for average match and nonmatch scores, as well as image quality scores between the left and the right irises. The results from ICE 2005 are publicly available [17].

The objective of ICE 2006, the second phase of the study, was to establish an independent performance benchmark for commercially available systems. Data was collected using the same protocol as ICE 2005 dataset, but under more challenging conditions. Data was collected from 240 subjects from University of Notre Dame on the LG 2200 camera with a modified quality control procedure, which resulted in about one-third of the iris images above the system defined as the quality threshold and the rest below the quality threshold. Eight groups agreed to participate in this study, but this time NIST conducted the matching operations in-house using all the algorithms. After the evaluation, only three out of the eight groups agreed to have their names and results published. The median FRR for the three systems was 0.012, 0.019, and 0.021 at an FAR of 0.001. The complete results from ICE 2006 are publicly available [18].

5.8.3 IRIS06

The IRIS06 was funded by the U.S. National Institute of Justice and DHS and was conducted by Authenti-Corp [19]. This study used three commercially available iris systems and collected data from 300 live subjects. The iris images were formatted using the ISO/IEC 19794-6 standard. This study reported acquisition error rates, FTE, FMR, FNMR, and transaction times for all three iris datasets and also concluded that left eye and right eye iris recognition performances are statistically similar. Table 5.4 provides a summary of results for the three systems.

A comparison of rates between the ITIRT and IRIS06 will show significant differences; remember the discussion from Chapter 2 on performance evaluation and how it is impacted by factors such as the number of individuals, training, the time period between successive transactions, and others.

Along with calculating the basic performance metrics for the three datasets, this study analyzed the following:

- *Interoperability of cameras:* Performance rates were better for native dataset comparisons than interoperable dataset comparisons where iris images from one camera were compared against iris images from another camera.

Table 5.4 IRIS06 Summary of Three Attempt Performances

Performance Metric	System A	System B	System C
FTA %	1.5	6.9	6.9
True match rate at 0.32 HD	99.7	97.3	99.4
FNMR = 1-TMR	0	1.8	0.4
Recognition transaction time	21.4	7.9	11.2

Source: [19].

- *Effect of eyeglasses:* Eyeglasses had a negative impact on recognition rates of datasets captured by two of the cameras.
- *Impact of habituation:* The time difference between enrollment and verification attempts, which ranged from 15 minutes to 8 weeks, did not have a significant impact on performance metrics.
- *Impact of yaw and roll:* Performance was not affected by the yaw and roll of the subject's face within a 20° range from neutral alignment.

5.8.4 NIST IREX

NIST established the Iris Exchange (IREX) umbrella project with the goal of supporting and increasing the adoption of iris image interoperability. IREX-I, the first subproject to be completed, evaluated the interoperability of three different iris image data formats, the impact of image compression on recognition performance, and the time to complete the system processes and to assess the ability of vendors to produce 19794-6:2005 conformant records. The ISO/IEC 19794-6:2005 standard specifies two different data formats discussed in Section 5.5. The IREX-I evaluated three new image data format proposals: the center cropped rectilinear format, the cropped and masked rectilinear format, and the unsegmented polar format, which are shown in Figure 5.5.

The results of the IREX-I, summarized next, were expected to be incorporated into the review of the ISO/IEC 19794-6:2005 standard. The results of IREX-I are publicly available [20].

- Use an uncropped and uncompressed rectilinear format where there are no image size constraints.
- Use a cropped rectilinear format when image acquisition software is capable of iris detection.

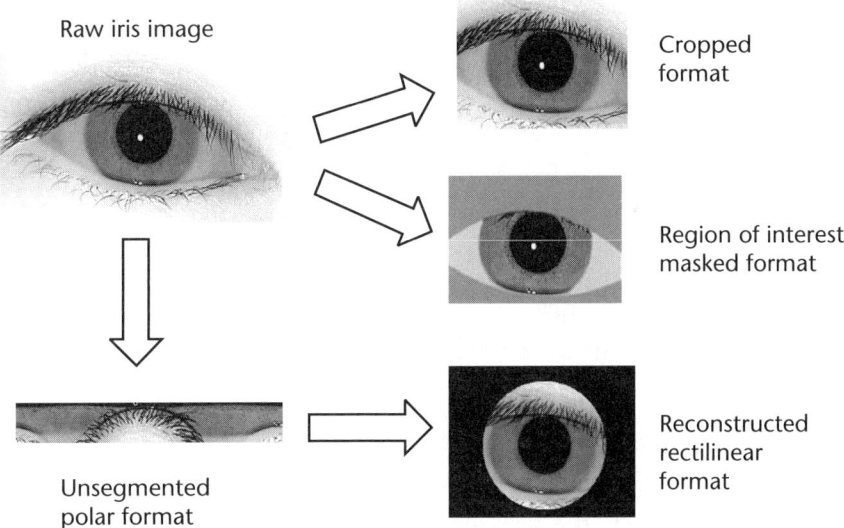

Figure 5.5 Iris image formats evaluated in IREX-I. (*Source:* NIST Interagency Reprot 7629, NIST, 2009. Reprinted with permission.)

- Use a cropped and masked rectilinear format for exchanging iris images of less than 3 KB.
- An unsegmented polar format is not recommended for interoperability purposes.

IREX-II Iris Quality Calibration and Evaluation (IQCE) is an ongoing project that aims to evaluate the predictability of recognition performance by iris image quality algorithms. The results from this study will provide a technical contribution to the ISO/IEC 29794-6 iris image quality standard. IREX-III is aimed at assessing the state of the art for large-scale iris identification systems where the number of records is likely to exceed 10^7. At the time of this writing, an IREX-IV was planned that would assess the verification performance of iris recognition algorithms for personal identity verification (PIV) cards.

The Noisy Iris Challenge Evaluation (NICE) is a public evaluation program conducted by the Soft Computing and Image Analysis Group of the University of Beira Interior in Portugal [21]. NICE.1, which concluded in 2008, evaluated the iris segmentation and noise reduction algorithms of all participants. The fundamental task of the algorithms participating in NICE.1 was to produce a segmented and enhanced binary iris image for all raw images provided in a test dataset. The NICE.1 organizing committee also created a database of optimally segmented and enhanced binary iris images for all test dataset images. The evaluation metric was based on the proportion of disagreeing pixels of a participant's output and the organizing committee's output. The rates were generated on a scale of [0, 1] where 0 indicated the lowest error rate possible and 1 indicated the highest error rate possible. The best performing algorithm in NICE.1 produced an overall classification error of 0.0131 [21]. A follow-up evaluation, NICE.II, concluded in 2010 and used a new dataset of visible light iris images captured in a noisy and on-the-move scenario. The evaluation protocol was modified to measure the disagreement of pixels between the participant's output and all available outputs provided by the NICE.II organizing committee. This resulted in a 1:many comparison and a decidability value was computed based on the interclass and intraclass disagreement values. The results from this evaluation are publicly available [22].

There are several databases that have been collected for testing and evaluation purposes. A detailed description of these is provided in Chapter 13.

5.9 Applications and Trends

Iris recognition has had quite a few successful deployments in large-scale applications in the last decade. It is even more noteworthy that these deployments have occurred in different countries, with goals as diverse as increasing convenience for frequent fliers to terrorist identification.

5.9.1 Privium System at Schipol Airport, Amsterdam

Launched in 2001, the Privium system is one of the longest-running iris recognition programs for border control for frequent fliers. This is an entirely voluntary, subscription-based system aimed at business travelers who do not want to spend time in immigration queues. On successful enrollment, the user is given a tamper-proof smartcard with his or her iris template stored on it. For authentication the user provides an image of his or her iris along with the smartcard to the system. The iris image is compared against the template stored on the smartcard and, based on the decision, the user is allowed to pass through or is sent to an immigration officer for manual inspection.

5.9.2 U.K. IRIS Program

The U.K. Iris Recognition Immigration Service (IRIS) has been in operation since 2005 and is a voluntary fast-track border-crossing service available to U.K. citizens. Users enroll both their iris images, along with their passport credentials, which are stored in a central database. Upon entering the United Kingdom, enrolled users go through special immigration lanes and provide their iris images to the system. If their irises successfully match, they are allowed through or else they have to be processed by an immigration officer. This is an identification system where the user makes no claim to an identity. There are more than 100,000 users in the system with over 10,000 transactions per week.

5.9.3 U.S. Department of Defense (DoD) Iris Recognition

The U.S. DoD has been a strong supporter of iris recognition technology in Iraq and Afghanistan and is primarily used for personnel screening and identifying individuals of interest in sensitive areas. Since 2003 the military has used mobile iris recognition as part of the Biometric Automated Toolset (BAT) to enroll and identify the indigenous population against a blacklist of individuals of interest. The military bases are supported by individuals who are members of the military and physical access is controlled using iris recognition. The use of this technology has expanded since it was deployed, and although there have been operational challenges such as harsh environment and user interface, the DoD considers this to be a valuable tool in their field operations.

5.9.4 United Arab Emirates (UAE) Expellee Program

The UAE rolled out its Expellee program in 2003 across all its airports, seaports, and land border points to identify individuals expelled from the UAE and attempting to reenter the UAE using false identification documents. This is by far the largest and most successful implementation of its kind and one that demonstrates true potential of this technology. When an individual is deported from the UAE, both of his or her iris images are enrolled and stored in a central database blacklist. Every individual entering the UAE through any of the airports, seaports, or land border points is identified against the blacklist. If the individual is matched against one of the expelled identities, an immigration officer conducts an investigation to determine the true identity of the individual. The blacklist contains over 1.5 million records and more than 6 billion comparisons are performed on a daily basis, with over 60,000 individuals identified correctly against the blacklist since the start of the program [23]. The system administrators claim a response time of less than 2 seconds for each comparison.

The United Nations High Commission for Refugees (UNHCR) has successfully used iris recognition to control fraud in the Afghanistan refugee repatriation program, which entitles refugees to financial, food, and shelter assistance. India initiated the AADHAR program in 2009 with the goal of providing a unique identity number to all residents. Both iris images, along with all 10 fingerprints, will be used to link the UID to the individual. The number of iris recognition systems is numerous and this trend is likely to continue, especially for large-scale systems.

5.10 Design and Deployment Considerations

5.10.1 User Interaction

The noncontact aspect of iris recognition places even greater significance on user interaction and training. Determining the depth of field is difficult for new users as there is no physical demarcation available. Most iris systems do not provide a visual of the iris image to the user, and it is difficult for users to determine how to improve their interaction so that they can provide a better iris image. Most of the first generation iris systems would only provide feedback to tell the users if they were within the field of view of the camera but not on what they needed to do to enter the field of view. The ITIRT and IRIS06 evaluation results indicated the need to improve user interaction so that the FTA and FTE rates can be reduced, along with improving throughput. The systems today are getting better at providing feedback that can guide the user into the appropriate position. System integrators have shown that outlining a capture box on the floor is an effective means of guiding the user. Some innovative systems show the entire face of the user as a visual feedback with an outline of two eyes. The users can position themselves in such a way that the outline is occupied with their own eyes.

User training can aid in removing initial doubts that a user may have about interacting with the system. A well-defined training procedure will improve the quality of the iris image, reduce the FTA and FTE rates, and increase throughput.

Mobile devices are being used in military applications currently and will definitely see growth in law enforcement and national ID programs. Use of these devices requires operator training, ensuring that the person operating the device can capture the best iris images in the least amount of time.

5.10.2 Medical Conditions

Medical conditions are an important factor to consider, especially when the system is designed to accommodate millions of users. Previous research has examined various ocular disorders and hypothesized about its impact on iris recognition, but very few have conducted experiments to test the hypotheses. Research conducted in 2004 on 55 patients who underwent cataract surgery concluded that only six patients could not be verified against their preoperative iris images [24]. The recommendation for real-world systems is to reenroll individuals who undergo cataract surgery. A more comprehensive study analyzing the impact of ocular pathologies on iris recognition was conducted by Aslam et al. [25]. Their experiment involved 54 patients who suffered from one of the following: glaucoma requiring laser iridotomy, inflammation of the iris, corneal pathologies, episcleritis, scleritis, and conjunctivitis. Their results indicated that, except for the inflammation of iris, the other eye pathologies had no significant impact on similarity scores, and all matching errors could be corrected by reenrolling the individuals. Glaucoma, which is a type of an ocular degenerative disorder, results in spots forming on the iris and thus interfering with the iris pattern. Aniridia is a condition in which individuals do not have an iris. This disorder affects about 1 in 100,000 people, which can be a significant number for extremely large-scale systems [9]. Nystagmus and strabismus are eye conditions that result in the involuntary movement of the eye and the inability to align both eyes, respectively. This can result in blurry or off-gaze iris images.

5.10.3 Nonideal Iris Images

Iris images that have a high degree of occlusion, unclear pupil and iris boundaries, or incorrect orientation could lead to lack of iris texture information for reliable matching. These can be detected using quality control algorithms, but future systems should also have the capability of recommending to the user how to avoid providing poor-quality images.

Colored and other types of cosmetic contact lenses can lead to false rejections because of the change in iris patterns. Clear contact lenses that completely overlap the iris do not have an impact on performance of the systems. The LASIK procedure used to correct vision is shown to have no impact on performance.

5.10.4 Environmental Effects

The amount of light entering the pupil affects its size and, in turn, the size of the iris. Low light conditions increase the size of the pupil, which could lead to a lack of iris texture information in the resulting image. Research has shown that iris images with a smaller pupil size perform better than iris images with larger pupils and that size difference in pupils between enrollment and recognition attempts can increase FNMR [26]. In such cases controlling lighting conditions in the deployment area can reduce these errors.

5.10.5 Interoperability

The results of ITIRT and IRIS06 evaluations highlighted the issues related to sensor interoperability. A large-scale implementation has to take into account these degradation issues and select the cameras and data formats that allow the highest degree of protection from a change in the technology. Currently, the safest choice is to select the same camera for enrollment and verification stations, but this could leave the system vulnerable to vendor tie-in.

5.10.6 Spoofing Attacks

The integrity of biometric data is essential to the overall security of an iris recognition system. First generation iris systems could be attacked using high-resolution printouts of iris images. Spectral analysis of live iris images and printed iris images exhibit differences that can be used for spoof detection [27]. Printed iris images cannot change their geometric proportions if the pupil contracts or dilates. The pupil size is affected by the impinging infrared light, which can be monitored to ensure liveness. Hemoglobin in the blood has specific absorption properties under infrared illumination. The vascular structure in the sclera can be used to detect if the eye being presented is live. Tracking a subject's response to blinking challenges initiated by the iris system is another simple but effective means of liveness detection. Spoofing techniques and spoof detection methods are discussed in Chapter 15.

5.11 Summary

Iris recognition has made great strides in the last decade, but it is far from being considered fully mature. The iris texture has shown enough distinctiveness for use in large-scale applications, but there is no empirical evidence regarding its stability. Researchers are currently exploring this question, and preliminary evidence points at measurable changes in similarity scores over a period of time [28]. The evaluation of iris images has shown that different color irises provide better-quality images in different wavelengths. Future iris systems could use multispectral imaging and choose the best-quality iris image. Iris segmentation still remains the most studied area of iris recognition, although research in user interface and iris capture at a distance will likely become as important in the near future. Iris recognition has already established itself as the biometric of choice for large-scale systems, and its proliferation will continue in the near future.

References

[1] Daugman, J., "Probing the Uniqueness and Randomness of IrisCodes: Results from 200 Billion Iris Pair Comparisons," *Proceedings of the IEEE*, Vol. 94, 2006, pp. 1927–1935.

[2] Bertillon, A., "La Couleur de l'iris," *Revue Scientifique*, Vol. 36, 1885, pp. 65–73.

[3] Safir, A., and L. Flom, "Iris Recognition System," U.S. Patent No. 4,641,349, February 1987.

[4] Johnston, R., *Can Iris Patterns Be Used to Identify People*, Los Alamos National Laboratory Technical Report, New Mexico, 1992.

[5] Daugman, J., "Biometric Personal Identification System Based on Iris Analysis," U.S. Patent No. 5291506, 1994.

[6] Matey, J. R., and L. R. Kennell, "Iris Recognition—Beyond One Meter," in *Handbook of Remote Biometrics*, M. Tistarelli, S. Li, and R. Chepella, (eds.), New York: Springer-Verlag, 2009, p. 37.

[7] Daugman, J., "New Methods in Iris Recognition," *IEEE Transactions on Systems, Man, and Cybernetics*, Vol. 37, 2007, pp. 1167–1175.

[8] Wildes, R., "Iris Recognition: An Emerging Biometric Technology," *Proc. IEEE*, Vol. 85, 1997, pp. 1348–1363.

[9] Bowyer, K. W., K. Hollingsworth, and P. J. Flynn, "Image Understanding for Iris Biometrics: A Survey," *Journal Computer Vision and Image Understanding*, Vol. 110, 2008, pp. 153–172.

[10] NIST, *IREX Overview,* NIST Report, 2009.

[11] Cukic, B., et al., "Image Quality Assessment for Iris Biometric," *SPIE Biometric Technology for Human Identification III,* Vol. 6202, 2006, pp. D1–D11.

[12] ISO/IEC, *ISO/IEC 19794-6:2005 Iris Image Data*, Geneva, Switzerland, 2005.

[13] Proena, H., and L. A. Alex, "Iris Recognition: An Analysis of the Aliasing Problem in the Iris Normalization Stage," *International Conference on Computational Intelligence and Security*, 2006, pp. 1771–1774.

[14] NIST, *American National Standard for Information Systems—Data Format for the Interchange of Fingerprint Facial, & Other Biometric Information—Part 1,* 2007.

[15] Orandi, S., and R. M. McCabe, *Mobile ID Device Best Practice Recommendation Version 1*, Gaithersburg, MD: National Institute of Standards and Technology, 2009.

[16] IBG, *Independent Testing of Iris Recognition Technology (ITIRT)—Final Report*, 2005.

[17] Phillips, P. J., et al., "The Iris Challenge Evaluation 2005," *IEEE Second International Conference on Biometrics: Theory, Applications and Systems*, Arlington, VA, 2008.

[18] Phillips, P. J., et al., "FRVT 2006 and ICE 2006 Large-Scale Experimental Results," *IEEE Transactions on Pattern Analysis and Machine Intelligence*, Vol. 32, May 2010, pp. 831–846.

[19] Authenti-Corp, *Iris Recognition Study (IRIS06)*, 2006.

[20] Grother, P., et al., *IREX I—Performance of Iris Recognition Algorithms on Standard Images*, Gaithersburg, MD, 2009.

[21] Proenca, H., and L. A. Alexandre, "The NICE.I: Noisy Iris Challenge Evaluation—Part I," *First IEEE International Conference on Biometrics: Theory, Applications, and Systems, 2007 (BTAS 2007)*, Crystal City, VA: IEEE Press, 2007, pp. 1–4.

[22] SOCIA, "NICE: II Noisy Iris Challenge Evaluation, Part II," 2010, http://nice2.di.ubi.pt/.

[23] Al-Raisi, A. N., and A. M. Al-Khouri, "Iris Recognition and the Challenge of Homeland and Border Control Security in UAE," *Telematics and Informatics*, Vol. 25, 2006, pp. 117–132.

[24] Roizenblatt, R., et al., "Iris Recognition as a Biometric Method After Cataract Surgery," *Biomedical Engineering Online*, Vol. 3, 2004.

[25] Aslam, T. M., S. Z. Tan, and B. Dhillon, "Iris Recognition in the Presence of Ocular Disease," *J. R. Soc. Interface*, Vol. 6, 2009, pp. 489–493.

[26] Hollingsworth, K., K. Bowyer, and P. Flynn, "Pupil Dilation Degrades Iris Biometric Performance," *Computer Vision and Image Understanding*, Vol. 113, January 2009, pp. 150–157.

[27] Daugman, J., "Demodulation by Complex-Valued Wavelets for Stochastic Pattern Recognition," *International Journal of Wavelets, Multi-Resolution and Information Processing*, Vol. 1, 2003, pp. 1–17.

[28] "Aging Irises Would Hobble Biometric Identity Checks Based on Iris Recognition," *Homeland Security Newswire*, 2010.

CHAPTER 6

Hand Geometry Recognition

The human hand is an extremely versatile organ that is used in virtually every aspect of our daily lives. The human hand exhibits physical characteristics such as the length, width, and thickness of fingers, the width of the back of the hand, and other features that are unique enough for the recognition of individuals. Contrary to popular misconception, hand geometry technology does not use fingerprint or palm print information; it focuses only on overall shape of the hand. Hand recognition is one of the older commercially available technologies and has been deployed in access control applications since the early 1970s. Current hand recognition systems are not suitable for use in identification applications and all commercial systems operate in verification mode (i.e., the user is required to provide a PIN, a username, or a card). Hand geometry recognition has several advantages:

1. The acquisition process is nonintrusive.
2. Feature extraction and matching are computationally efficient.
3. There is less reluctance towards using hand geometry [1].
4. Cuts and abrasions on the hand do not affect this technology.

This chapter will discuss the history, underlying technology, applications, and deployment considerations of this technology.

6.1 History

Archeologists have discovered prehistoric carvings that included a human hand impression and conjectured that the impression was used to link the artist with the art. In the 1960s while working for his clients Robert Miller observed that their hand sizes were different and started working on a device to verify an individual's identity based on his or her finger lengths. He was granted a patent in 1971 based on a mechanical system that measured the length of a person's fingers to recognize an individual [2]. This concept was commercialized by Identimation Corporation, which created an electronic device for capturing the length of fingers using an overhead illumination source and a flatbed of photocells. The area of the photocells covered by the hand has less electricity passing through it and the difference in electricity is measured to create the contour of the hand. In 1988 David Sidlauskas was granted a patent on a hand recognition system that took advantage of low-cost digital imaging and signal processing, and Recognition Systems, Inc., (RSI), was incorporated to commercialize this technology [3]. The original RSI systems captured the dorsal side and a profile view of a person's right hand up to his or her wrist, and this still currently serves as a basis for their systems. RSI products have gone through four generations of advances and today are the largest vendor of this technology. The principle of hand geometry has also been applied to finger shape recognition. Commercial systems based on geometry of the hand typically operate in verification mode. Figure 6.1 shows an image of a commercial hand geometry system.

6.2 Image Acquisition

All commercially available technologies use a camera for capturing the hand image, and a few research systems have experimented with flatbed scanners. Most hand geometry systems provide a flat surface, also called a platen, for the user to place his or her hand. Commercial systems use a configuration of pegs on the platen that guides the hand placement of the user. These pegs have several functions:

1. The pegs force the user to spread their fingers, which ensures a clear demarcation of edges in the resulting image.
2. The pegs provide reference points for feature extraction and ensure that the hand is placed in a consistent orientation, which reduces feature extraction complexity.
3. The correct placement of the hand around the pegs, along with pressure applied by fingers on the pegs, triggers the image capture mechanism. This reduces acquisition errors because of incorrect hand placements.
4. The pegs provide a degree of liveness detection as the hand being placed has to be live in order to squeeze the pins.

Infrared illumination is used in commercial hand recognition systems as a grayscale image simplifies the feature extraction process. Both the dorsal and the ventral

HandPunch® GT400

Figure 6.1 Hand geometry system. (*Source:* Ingersoll Rand Security Technologies. Reprinted with permission.)

sides provide required information for feature extraction, but commercial systems concentrate on the dorsal view, along with a side view of the hand. Orthographic scanning is used in commercial systems to capture multiple 2-D perspectives of the hand to infer 3-D measurements of the hand. The resolution of cameras used in hand recognition is much lower compared to other image-based biometric systems, typically in the range of 100–200 dpi. Research has shown that images collected at even 30 dpi showed only a minor degradation in [4]. Due to the low resolution of the sensor, the resulting image is not affected by minor variations in the shape of the hand or in the placement of the hand. The optical path, which is the distance between the hand and camera, is required to be at least 10 inches to capture the entire hand. The physical dimensions of such a sensor would be too large for most practical systems. Commercial systems overcome this problem by using an array of mirrors that reduces the physical dimensions of the optical configuration.

6.3 Feature Extraction

The first step of the hand segmentation process is the conversion of the input image into a grayscale image, which is then converted into a binarized image. The binarized image is a silhouette of the hand where the background and the hand area are in contrasting colors, as shown in Figure 6.2. A filtering process is applied if any imaging artifacts such as hotspots, uneven reflection, or other noise are present in the hand region of the image. The next step is to trace the border of the hand silhouette. Although this seems like a relatively easy process, the edge detection algorithm has to be extremely precise and also compensate for any breaks in the image due to rings or other hand jewelry. Edge detection algorithms based on watershed transforms, Canny edge operators, and the Sobel function have been used in hand recognition research with varying degrees of success.

Normalization of the hand silhouette is necessary to compensate for any variability in hand placement and finger placement. The orientation of the overall hand shape and the direction of fingers are aligned based on some standardized directions and size. The alignment process has to ensure that distortion is not introduced in the resulting aligned image. Human fingers are flexible and have a wide range of motion around several pivot points such as the connection of proximal phalanx

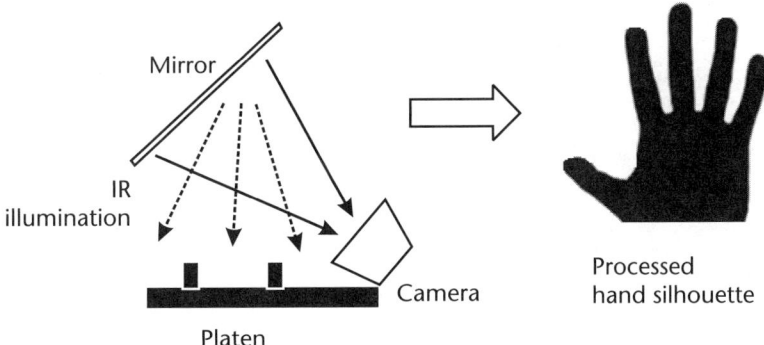

Figure 6.2 Imaging system.

with metacarpals. This makes it difficult to create a global alignment method. The alignment process uses certain properties of the underlying skeletal structure of the human hand. The dorsal side of the palm is a rigid area that can only rotate around the wrist, while the fingers can pivot at the point where the proximal phalanx of each finger meets the metacarpals, or the knuckle region. These four points, corresponding to four fingers, are identified as pivot points. A horizontal axis is created that passes through these pivot points and the angle of orientation of each finger is calculated and stored. The pivot point information and angle parameters are stored with the enrollment templates and are used for every subsequent input sample that is to be compared against the template. Normalization is essential in systems that do not use pegs or platens to compensate for the rotation and depth of view of the imaging system. Typically, the region below the thumb up to the wrist is de-emphasized in hand recognition because of the uncertainty in detecting boundaries in the region. For example, clothing that extends beyond the wrist or improper placement of the hand can introduce a false boundary in the resulting image.

The hand region data can be represented using two different methods: geometric features and contour features. Geometric features of the hand, such as widths of fingers, height of fingers from a reference point, thickness of fingers, width of the palm, surface area of the different parts of the hand, and angles between finger directions, are extracted from the hand silhouette. The coordinates of the fingertips and the base of the fingers are generally used as anchor points for calculating the distances, as shown in Figure 6.3. The resulting measurements are used to create a feature vector for either enrollment or verification.

Contour features use edge mapping techniques such as the Freeman chain code (FCC) to store the outline of the hand as a closed object. This is an effective data compression technique for systems that want to store the global shape of the hand. The FCC is a sequence of *n* links that connects contour pixels of the object at predetermined intervals. The direction of each link can be shown based on a predetermined number of directions in a 360° space. An eight-directional space is typically used in determining the FCC of an object, as shown in Figure 6.4.

The hand segmentation process in commercial systems is simplified by the use of a reflective platen and infrared illumination. Segmentation is more complex for color images than grayscale images; infrared illumination controlled by the system reduces any blurs and hotspots that can interfere with the segmentation process.

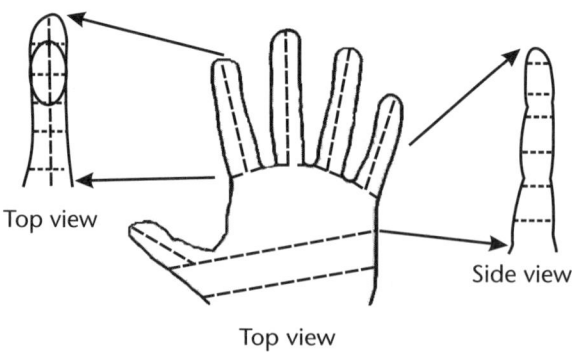

Figure 6.3 Geometric features.

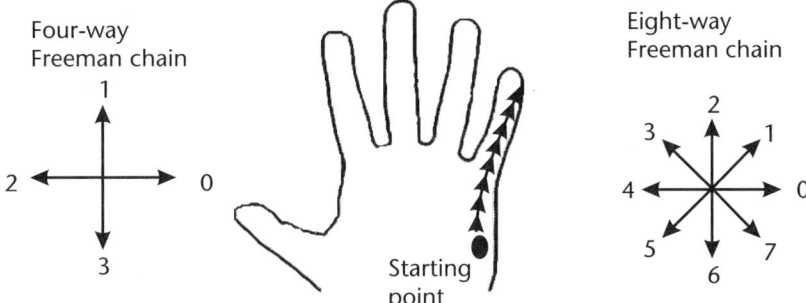

Figure 6.4 Contour features: the Freeman chain code.

The platen in commercial systems is coated with reflective material that increases the contrast between the background and the hand image. The system designed by RSI uses about 90 different geometric features to produce an enrollment template of approximately 18 bytes.

6.4 Feature Matching

Different algorithms for matching feature vectors have been explored, including the Euclidean and Mahalanobis distances, the Hamming distance, Gaussian mixture models (GMM), and neural networks [5]. In this specific research stochastic models based on GMM provided the best performance results, though at a higher computational cost and larger template size.

The Hausdorff distance metric has been used for comparing hand silhouettes or contours. The contours of the hand image can be viewed as a set of points, and this algorithm calculates a distance metric between two sets. This algorithm is often used in computer vision for comparing binarized images. A modified version of the algorithm that is robust to minor variations has been used in research experiments [6, 7].

Data reduction techniques such as principal component analysis (PCA) and independent component analysis (ICA) are applied to geometric features as well as the hand silhouettes. These methods extract statistically independent components from the signal without losing information that contributes to its uniqueness. In such methods the difference in principal components is used for comparing two feature vectors.

The size and shape of a human hand undergo minor variations due to physiological processes such as weight gain and loss of moisture. Template adaptation is used by hand geometry systems to accommodate these physical changes. Most systems will update the enrollment template with the changes upon a successful verification. Any major changes to the shape or size of the hand, such as the presence of a large ring on a finger, cannot be addressed by template adaptation. Either a reenrollment or a removal of physical objects such as the ring is required.

6.5 Performance Evaluations

Hand geometry recognition is currently limited to private industry with no apparent use in the law enforcement domain, and thus has not been subjected to many large-scale studies such as fingerprint, face, and iris recognition. In 1991 Sandia National Labs tested the ID-3D system from RSI. Data was collected from around 100 people over a course of 3 months, resulting in more than 5,000 genuine and imposter transactions [8]. The equal error rate (EER) for one-attempt, two-attempt, and three-attempt verification transactions were all less than 0.2% at the threshold recommended by the vendor.

The National Physical Laboratory (NPL) collected hand geometry images in a test conducted in 2001. Approximately 200 subjects participated with at least a 1-month gap between enrollment and verification transactions. The FTA and FTE were 0% for both enrollment and verification transactions [9]. A three-attempt EER of 0.5% was calculated for this study.

In 2003 the International Biometric Group (IBG) conducted round 5 of Comparative Biometric Testing (CBT), which included a hand recognition system from RSI. However, the results for this test are not publicly available. There is a significant body of academic research in the field of hand geometry recognition. Most of this research has explored different methods for extracting and matching features rather than evaluating the current state of the technology. Table 6.1 summarizes the results from a selected list of hand geometry research studies [10–15].

6.6 Standards

The current international standard for hand geometry data interchange, ISO/IEC 19794-10, was published in 2007 [16]. The basic features of the hand are encoded using four-way or eight-way body-centric Freeman chain coding schemes. The edge of the hand is traced and the direction of the contour is recorded using the specified Freeman chain coding at several points along the outline. Other data describing the hand the data capture system are also included in the standard, which is described in Table 6.2.

The U.S. standard, ANSI/INCITS 396-2005, has been withdrawn in favor of the ISO/IEC standard.

Table 6.1 Summary of Hand Geometry Experiments

Reference	Number of Subjects	Features	Matching	Performance Metrics
[11]	100	Geometry	Bayes classifier	EER = 0.0012
[12]	50	Geometry	Mahalanobis distance	FAR = 0.01, FRR = 0.17
[13]	53	Contour	Alignment of shape	FAR = 0.01, FRR = 0.06
[14]	96	Geometry	Distance metric	EER = 0.03
[7]	458	Contour	Modified Hausdorff distance	EER = 0.02
[15]	470	Geometry	Sum of feature deviation	FAR = 0.0045, FRR = 0.034

Source: [10].

Table 6.2 Overview of ISO/IEC 19794-10

Parameter	Description
Hand integrity	Indicates any problems with the captured image such as a missing finger
Hand identifier	Indicates the hand and the angle from which it was captured
Quality	Metric describing contour edge determination
Distortion	Displacement of the image from the paraxial position
Camera position	3-D coordinate position of the camera
Target position	3-D coordinate position of the hand
Silhouette starting point	2-D coordinates of the starting point for the Freeman chain coding
Freeman chain coding	Identify if four-way or eight-way coding is used
Scanning technology	Type of acquisition system: optical camera, scanner, or unspecified

Source: [16].

6.7 Applications and Trends

Hand geometry recognition is predominantly used for physical access control and time and attendance applications. The rugged nature of the device, its resilience towards extreme weather conditions, the robust performance in verification mode, and the lack of inherent objections from users in using their hands for recognition purposes have made it an appealing technology for physical access control. The cost-benefit analysis of replacing traditional lock-and-key mechanisms with hand geometry technology is easier to quantify in numerical terms, which is one of its biggest advantages.

The University of Georgia deployed hand geometry recognition for physical access control in 1972, and it is the longest running system of its kind in the United States [17]. Since then, the hand geometry recognition system has been expanded to the administration of the student meal program and access to residence halls, sports recreation facilities, and other controlled areas. The entire student population is enrolled in the system, allowing the university administration to streamline its access control infrastructure and provide convenience to students.

Since 1991, the San Francisco International Airport has been using hand geometry recognition for secure access to its airport operations areas such as taxi ways, jet ways, and the tarmac. All doors throughout the entire airport are equipped with 300 hand readers, and more than 18,000 employees use this system for access control.

In the 1996 Olympics hosted in Atlanta, Georgia, hand geometry recognition was used for entry to the Olympic Village. All authorized individuals were provided with a smartcard that contained their enrollment template. The sample captured from the user was then compared to the enrollment template on the smartcard during the verification process.

Some other prominent deployments are:

- Ben Gurion airport in Israel uses hand geometry to process trusted travelers through express security lanes.

- Portos Santos in Sao Paulo, Brazil, uses hand geometry for the authentication of 75,000 employees in the port area.
- All operational nuclear power plants in the United States use hand geometry recognition for entry to the physical premises.

The proliferation of mobile devices equipped with cameras has opened up the possibility of hand recognition using such devices. Researchers investigated hand recognition using a color image of 72 dpi captured using a mobile phone camera [18]. The images were taken in visible illumination without a platen or pegs and an EER of 3.7% was observed on a dataset of 45 users. Although this is an experimental system, it shows potential for use in consumer-facing applications.

6.8 Design and Deployment Considerations

The physical size of a hand geometry reader, its ability to operate only in verification mode, and the use of a single hand are some of the limitations of this technology. This section also discusses operational challenges that arise when the technology is deployed in real-world applications.

6.8.1 Environmental Issues

The hand geometry scanner works well in outdoor conditions and is the technology of choice for physical access control. However, there are some extreme conditions that can interfere with the imaging mechanism. Hand geometry imaging uses infrared illumination and an array of mirrors. The infrared waves from direct sunlight incident on the device platen can interfere with internal illumination. The reflection of light off the mirrors can also lead to the blinding of the camera and subsequent acquisition errors.

Extreme cold weather can also affect image capture. When the human hand comes in contact with a cold surface, the thermal transfer generates vapors from the hand and creates spurious edges in the hand contour. This issue is resolved by keeping the hand geometry reader sufficiently warm using a heater. The physical placement and internal heating mechanism for the hand reader to address environmental issues are important operational considerations.

6.8.2 User Interaction

Hand geometry recognition requires physical contact between the user and the capture device that leads to user interaction issues. Current commercially available hand readers are designed for right-handed individuals and have guidance pegs. The hand needs to be placed correctly around the guidance pegs in order to trigger the capture mechanism. Although this is an effective antispoofing technique, it can create problems for new users. An incorrect placement will not trigger the capture mechanism and cause inconvenience to the user. Supervised training for new users is a recommended practice to reduce FTE and FTA, which can lead to an additional cost. Empirical results of the right-handed capture setup have not shown an adverse impact on performance, but the level of effort for individuals who are naturally left-handed will be higher compared to right-handed individuals.

The height of the hand reader is another important consideration. Research has shown that the height of the sensor has an impact on performance in biometric technologies that require physical interaction [19]. If the average height of the target population is known ahead of the implementation, it should be taken into account for the placement of the sensor.

Missing fingers or medical conditions such as arthritis might not allow a user to enroll in the system. Research has indicated that individuals who were not able to fully extend their little fingers and straighten out their hands proved to be problematic during verification [1].

6.8.3 Other Performance Considerations

Users who wear large rings or other jewelry on their fingers or hands can pose a challenge to the hand geometry system as it leads to a fusion of adjacent fingers in the silhouette data. Typically, users who wear rings during enrollment and verification do not create problems for the system. Users who either remove a ring or start wearing a ring between enrollment and verification have a higher probability of being falsely rejected because of change in the hand contours. An administrative procedure that asks users to remove rings when interacting with the system can mitigate this issue.

User clothing such as long sleeves of shirts and jackets that extends beyond the wrist line can interfere with the segmentation process. User training and education are effective solutions to this problem.

The human hand size is affected by moisture content in the body. This is an important factor to consider where the target population is expected to have a high level of variability in body water content based on climate conditions or their type of work.

Certain types of tattoo inks have an absorption index that can interfere with infrared illumination. If a user has a tattoo on the dorsal side of his or her hand, it could lead to problems during the segmentation process.

Researchers have also carried out spoofing attacks on hand recognition devices, which is discussed in detail in Chapter 15.

6.9 Summary

Hand geometry recognition is a proven technology with more than two decades of real-world implementations. The biggest limitation of the technology is that it works only in verification mode and thus cannot be deployed in identification scenarios. Academic research conducted on peg-free platens and platen-free systems has shown varying levels of success [20]. Such systems represent the future direction of this technology. The human hand provides several biometric features and this technology offers several multimodal fusion opportunities. Preliminary research has explored the use of texture information from the hand that can improve overall performance of the system. The advantages of this technology in time and attendance and physical access applications will ensure that it remains relevant in the near future and become more tightly integrated into enterprise security infrastructures.

References

[1] Kukula, E., and S. Elliott, "Implementation of Hand Geometry: An Analysis of User Perspectives and System Performance," *IEEE Aerospace and Electronic Systems Magazine*, Vol. 21, March 2006, pp. 3–9.

[2] Miller, R. P., "Finger Dimension Comparison Identification System," U.S. Patent No. 3576538, 1971.

[3] Sidlauskas, D., "3D Hand Profile Identification Apparatus," U.S. Patent No. 4736203, 1988.

[4] Dutağacı, H., B. Sankur, and E. Yörük, "A Comparative Analysis of Global Hand Appearance-Based Person Recognition," *Journal of Electronic Imaging*, Vol. 17, 2008.

[5] Sanchez-Reillo, R., C. Sanchez-Avila, and A. Gonzalez-Marcos, "Biometric Identification Through Hand Geometry Measurements," *IEEE Transactions on Pattern Analysis and Machine Intelligence*, Vol. 22, 2000, pp. 1168–1171.

[6] Takacs, B., "Comparing Face Images Using the Modified Hausdorff Distance," *Pattern Recognition*, Vol. 31, 1998, pp. 1873–1881.

[7] Yoruk, E., et al., "Shape-Based Hand Recognition," *IEEE Transactions on Image Processing*, Vol. 15, July 2006, pp. 1803–1815.

[8] Holmes, J., L. Wright, and R. Maxwell, *A Performance Evaluation of Biometric Identification Devices*, Sandia National Laboratories, Albuquerque, NM, 1991.

[9] Mansfield, T., et al., *Biometric Product Testing Final Report*, Teddington, U.K., NPL, 2001.

[10] Duta, N., "A Survey of Biometric Technology Based on Hand Shape," *Pattern Recognition*, Vol. 42, November 2009, pp. 2797–2806.

[11] Golfarelli, M., "On the Error-Reject Trade-Off in Biometric Verification Systems," *IEEE Transactions on Pattern Analysis and Matching*, Vol. 19, 1997, pp. 786–796.

[12] Jain, A. K., "A Prototype Hand Geometry Based Verification System," *Proceedings of Second International Conference on Audio and Video-Based Biometric Person Authentication (AVBPA)*, Washington, D.C., 1999, pp. 166–171.

[13] Jain, A. K., "Deformable Matching of Hand Shapes for User Verification," *Proceedings of the IEEE International Conference on Image Processing (ICIP)*, Kobe, Japan, 1999, pp. 857–861.

[14] Covavisaruch, N., et al., "Personal Verification and Identification Using Hand Geometry," *ECTI Transactions on Computer and Information Technology*, Vol. 1, 2005, pp. 134–139.

[15] Adan, M., et al., "Biometric Verification/Identification Based on Hands Natural Layout," *Image and Vision Computing*, Vol. 26, April 2008, pp. 451–465.

[16] ISO/IEC, *ISO/IEC 19794-10:2007 Hand Geometry Silhouette Data*, Geneva, Switzerland, 2007.

[17] UGA, "UGA Safe and Secure," 2010, http://www.uga.edu/safeandsecure/.

[18] de Santos Sierra, A., et al., "Silhouette-Based Hand Recognition in Mobile Devices," *IEEE 43rd Annual 2009 International Carnahan Conference on Security Technology*, 2009.

[19] Theofanos, M., et al., *Effects of Scanner Height on Fingerprint Capture*, National Institue of Standards and Technology, 2006, p. 58.

[20] Amayeh, G., et al., "Peg-Free Hand Shape Verification Using High Order Zernike Moments," *Computer Vision and Pattern Recognition Workshop*, New York, 2006.

CHAPTER 7

Speaker Recognition

Speaker recognition, also known as speaker biometrics or voice recognition, uses distinct characteristics of a person's voice for recognition. Speaker recognition is based on the premise that physical and behavioral components responsible for voice generation are unique to the individual, and therefore the resulting voice signal also has unique characteristics. The vocal cavities such as the nose and the throat and the size and shape of vocal chords are physical features that are presumed to be unique to an individual. The movement of the jaw and lips, an acquired accent, and environmental influences are behavioral components that are also not replicable for two different individuals. The voice of a person thus is a combination of several unique physical and behavioral factors that have proven to be relatively unique by more than 30 years of research. Automated speech recognition (ASR), which is a related field of study, is often confused with speaker recognition. ASR is used for converting voice into corresponding text and involves the process of transcription, whereas speaker recognition is used for authenticating a user. ASR is not concerned with who is speaking, but rather with deciphering what is being said. Speaker recognition is a compelling biometric technology for the following reasons:

1. Voice is a natural signal that humans learn to produce intuitively.
2. Voice recognition is used implicitly in our daily lives.
3. The ubiquity of telecommunication infrastructure reduces the need for specialized hardware.

Speaker recognition is an active area of research with great commercial potential, but until now the real-world adoption of this technology has been limited. Speaker verification systems outperform speaker identification systems, but both are supported by a substantial body of research. This chapter discusses the current state of technology, successful deployments, and the operational challenges facing commercial deployment speaker recognition.

7.1 Generating Voice

Speech is produced using the vocal box, vocal chords, which are a pair of thin, long muscles, and the oral and nasal cavities. The process of breathing pulls in air through the nose and mouth and pushes it down the larynx and trachea and then into the lungs. This flow is reversed when air is breathed out. The vocal chords, located in the trachea, vibrate when the airstream passes through them and produce sound as the air moves through the larynx. The sound can be manipulated by changing the length and tension of the vocal chords. After the airstream passes through the vocal box, it enters the vocal tract comprised of oral and nasal cavities. The size, shape, and position of these cavities add resonance to the partially created sound. Finally, the interaction of articulators, which include lips, teeth, tongue, soft palate, and jaw muscles, generates intelligible speech.

7.2 History

During the first half of the twentieth century, researchers at Bell Labs discovered the relationship between intensity variations in sound frequencies and its sound characteristics called *spectrograms* [1]. During World War II humans were trained to compare spectrograms visually and recognize certain words and speakers. In 1962 Lawrence Kersta from Bell Labs coined the term *voiceprints* and explored the possibility of voiceprint comparisons [2]. As computing technology and understanding of voice signals advanced, automated recognition of speakers became the new goal. Bogert et al. introduced the concept of *cepstrum* (pronounced kep-strum) coefficients, which has laid the foundation of the majority of speaker recognition systems [3]. Figure 7.1 shows a voiceprint that maps the frequency of the signal against time. In the mid-1980s NIST created the Speech Group with the aim of studying and furthering the current state of the art in speech and speaker recognition. Since then NIST has begun several initiatives to facilitate advancement of speech processing techniques, which are discussed in Section 7.8.

7.3 Speaker Recognition Systems

Speaker recognition is performed in both identification and verification modes. Speaker recognition research has deep roots in pattern recognition research and academia literature will often use the terms "training" and "testing" to describe the process of enrollment and recognition, respectively. Speaker recognition systems are categorized based on types of speech samples that a user is required to provide during enrollment and recognition stages (Figure 7.2): *text-dependent*, *text-prompted*, and *text-independent*.

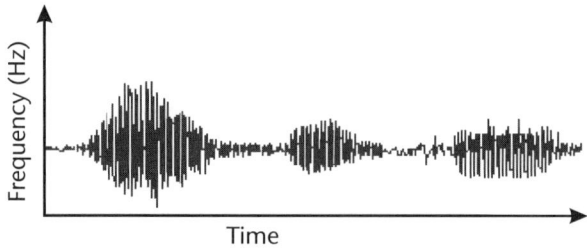

Figure 7.1 Spectrogram.

7.3 Speaker Recognition Systems

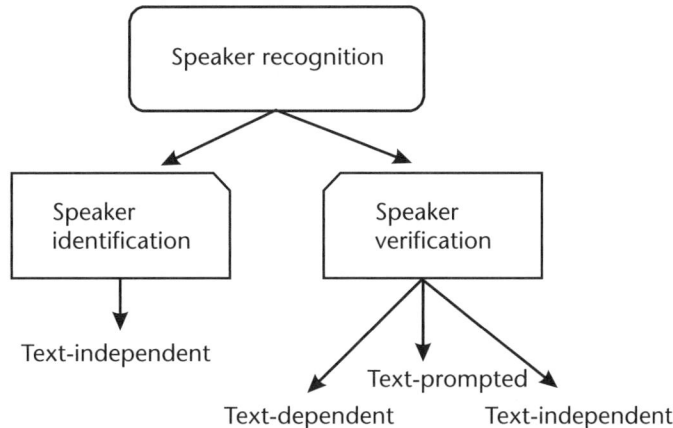

Figure 7.2 Types of speaker recognition systems.

7.3.1 Text-Dependent

Text-dependent systems have explicit knowledge of the words or phrases that the user will speak and the order in which they will be spoken, which remains the same for every transaction. This is a predetermined word, account number, or passphrase that is used during enrollment. Most commercially available systems fall into this category as they more resilient to environmental issues and generate fewer errors.

7.3.2 Text-Prompted

Text-prompted systems direct the user to speak a specific word or phrase, but the order in which it is spoken is always different. This is an extension of a text-dependent system because the system has prior knowledge of the possible words that can be spoken by the user. Such a system also requires the user to enroll a larger speech sample to ensure that every possible speech component is recorded. This type of a system provides better protection against replay attacks, which use recorded voice samples. The use of text-prompted systems is restricted to consumer-facing applications.

7.3.3 Text-Independent

In text-independent systems the user is not constrained to a set of specific words or phrases. The system has no advance knowledge of the speech sample—the enrollment and recognition samples could be completely different. This is a much more difficult problem to solve as it requires matching features of two samples that might not have a single syllable in common. Text-independent systems have traditionally garnered more interest from law enforcement and intelligence agencies for identifying persons of interest using telecommunication signals. These systems can be used for speaker identification (i.e., performing matching without a claim to an identity), but error rates are typically higher than 10% depending on the level of background noise.

7.4 Information Levels

A person's speech contains vast amount of information that can be categorized into four different levels based on the granularity of information in speech. These levels are explained graphically in Figure 7.3.

7.4.1 Idiolectal

This is the highest level of information in a person's voice and refers to a person's specific speech idiosyncrasies and habits. Humans are generally better at telling apart familiar people by hearing them compared to unfamiliar people. A person's word usage is influenced by his or her surroundings and other geocultural factors and provides a level of discrimination from other individuals [4].

7.4.2 Phonotactics

This refers to the phonetic sequences that are present in the utterance of a word or phrase. A person's speech is broken into tokens, or phones, which can then be transcribed using some form of a standard 1:1 mapping to word symbols. This information level is also influenced mainly by behavioral and geocultural factors.

7.4.3 Prosody

These features describe intonation, stress, rhythm, pitch, and tone of sounds that comprise a person's voice. By changing the prosodics of spoken language, the intent and significance of what is being spoken are affected. Prosody gives clues about the emotional state of the speaker, whether a question is being asked, which words are being emphasized for overall effect, and so forth.

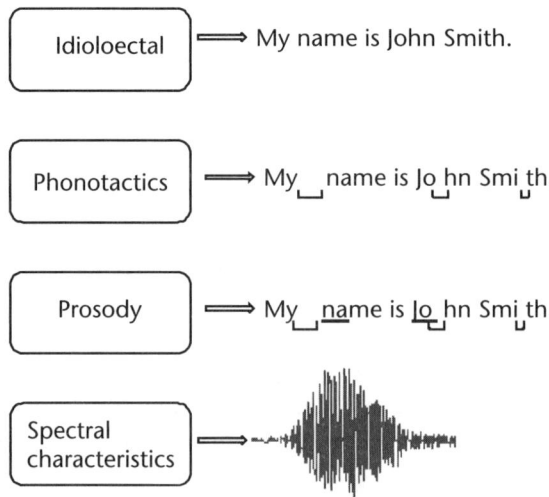

Figure 7.3 Speech information levels.

7.4.4 Spectral Characteristics

This information level describes the frequency and acoustic information of speech. For over four decades, spectral characteristics have constituted the majority of research in speaker recognition and form the basis for most commercially available systems. The majority of research in speaker recognition is based on extracting and matching spectral characteristics, although idiolectal, phonotactics, and prosody information levels have been used to supplement the uniqueness of the user. This chapter focuses on underlying technology, standards, and operational challenges as they relate to spectral characteristics.

7.5 Feature Extraction

Converting high-level audio signals to a set of compact, numerical feature vectors is done in a number of different ways, but most speaker recognition systems rely on the *cepstral* representation of the speech. The word cepstral is derived by reversing the first half of the word spectral and refers to the manipulation of spectral characteristics. The cepstrum is a common transformation technique used to separate the signal that contains words and pitch from the transfer function (spectral envelope), which represents the quality of the signal. Readers interested in the algorithm details of cepstrum extraction should read Bimbot et al. [5]. The two most popular techniques of representing cepstral parameters are Mel frequency cepstral coefficients (MFCC) and linear prediction coding (LPC). The process of signal enhancement and windowing, common to MFCC and LPC cepstral parameters, is described next.

7.5.1 Signal Enhancement

The goal of this phase is to enhance high frequencies of the spectrum that can be reduced by channel effects, the transfer function of microphones, and the voice production process in general.

7.5.1.1 Windowing

The voice signal is quasi-stationary (i.e., it varies slowly with time). When the voice signal is analyzed over a small period of time, empirically shown to be around 20–30 milliseconds, the signal appears stationary. The goal of the windowing operation is to divide the continuous voice signal into windows of a predetermined time period so that the voice signal is close to stationary in that window (see Figure 7.4).

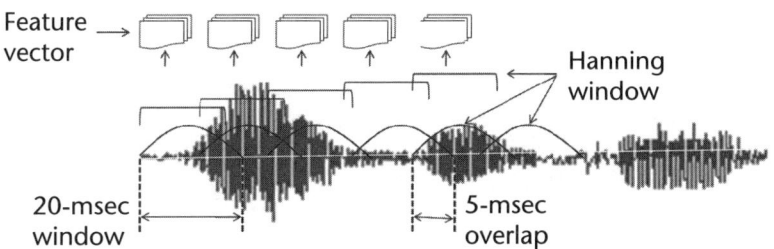

Figure 7.4 Windowing illustration.

This window is shifted over the entire voice signal and broken down into multiple frames that are analyzed one at a time. The shift distance is lesser than the size of the window, which results in an area of overlap between successive windows. Empirically a shift distance of 5–10 milliseconds has been used successfully in speaker recognition systems.

A rectangular window would lead to a total overlap of the signal in the common area and result in side effects in the final feature vectors of the both windows. In order to minimize this, Hamming or Hanning windows are used as they taper the signal to the sides and cause lesser spectral distortion.

7.5.2 Mel Frequency Cepstral Coefficients (MFCC)

A fast Fourier transform (FFT), which converts the signal from the time domain into frequency domain, is calculated for each window. The output of this phase is also called the spectrum of the signal. Readers interested in implementations of FFT should read Oppenheim et al. [6].

MFCC processing is based on how human ears react to audio signals; human perception of the frequency of sound does not follow a linear scale. The Mel frequency scale is similar to the human auditory scale and is linearly spaced below 1 kHz and logarithmically above 1 kHz. A collection, or bank, of Mel scale frequencies is multiplied one by one with the spectrum to calculate an average value in a particular frequency band. The cepstral coefficients are calculated by converting the Mel spectrum back to the time domain and this process is repeated for each window. Readers interested in the algorithm details of MFCC should read [7].

7.5.3 LPC Cepstral Parameters

LPC is another popular technique used in speaker recognition and is based on the linear model of speech production. Speech, as discussed earlier, is a result of various organs such as vocal chords, vocal tract, nasal tract, and oral cavity. Each of these sources is represented by a filter of different energy. It is possible to represent speech signal at a high level with an auto regressive moving average (ARMA) filter. The coefficients of the ARMA filter are calculated for each window that forms the LPC coefficients. These LPC coefficients are then converted into cepstral coefficients. Readers interested in algorithm details of LPC cepstral parameters should read [7].

Once the feature vectors are created using either MFCC or LPC cepstral parameters technique, the next step is to determine which vectors correspond to the audio signal of interest and which vectors belong to silent pauses or background noise. The feature vectors of interest are retained and the rest are discarded.

7.6 Feature Matching

The cepstral parameters generated during the feature extraction process can be matched using various techniques that are discussed in this section. The various algorithms and their uses are shown in Table 7.1.

7.6 Feature Matching

Table 7.1 Algorithm Classification

Text-Dependent	Text-Independent
DTW	GMM
HMM	SVM
	Vector quantization

7.6.1 Distance-Based Methods

This technique is predominantly used in text-dependent and text-prompted systems. The template consists of a series of feature vectors generated from the user's audio signal. The enrollment template is compared to the verification attempt and a similarity score is calculated between the two sets of feature vectors based on the distance between them. The similarity score then is compared to a threshold value and a determination is made about the identity claim. Dynamic time warping (DTW) is a popular method used to align the enrollment and verification templates as the two audio signals will not have the same speech rate. Nearest neighbor (NN) methods have also shown good results but are computationally intensive. In this method a distance metric is calculated between each window of the enrollment template and verification template. The minimum NN distances for each window are averaged to generate a similarity score. This similarity score is compared to a threshold value and a determination is made about the identity claim.

7.6.2 Model-Based Methods

In text-prompted and text-independent systems it is not feasible to create feature vectors for every possible occurrence in a voice signal. In such cases stochastic models are used for matching two voice signals. Stochastic models formulate the pattern matching problem as a hypothesis test where

H_0: Sample X is from speaker S

H_1: Sample X is not from speaker S

The likelihood ratio (LR) test described in (7.1) is used to compare how well the enrollment and verification models fit. If the probability is greater than a threshold, then a match is declared; otherwise, a nonmatch is declared.

$$LR = p(X|H_0)/p(X|H_1) \tag{7.1}$$

The effectiveness of the LR test depends on appropriate selection of the likelihood calculation function. Empirically, Gaussian mixture models (GMM) have provided the best results for text-independent systems. Readers interested in GMM implementation in speaker recognition are encouraged to read [8]. The LR test also requires a model for the alternative hypothesis. Universal background models (UBM) are a popular method for representing all possible alternatives for the hypothesized speaker. UBM is generated from voice samples that represent features from the general population of users and conditions, but there are no individual specific features. For example, consider a speaker verification system that will be used by female users over the age of 25 in an office environment. In such a case the

UBM consists of voice samples from a large set of female users, over the age of 25 and speaking in an office environment. This subpopulation is used as the alternative hypothesis for the LR test.

Hidden Markov models (HMM) have the ability to incorporate temporal information available in text-dependent and text-prompted systems. A first-order Markov model is a finite set of states with a probability distribution function that dictates state transitions. HMMs encode statistical variations of the temporal features of an audio signal. During enrollment HMM parameters are estimated and stored for the user. During verification the likelihood of input sequences is computed against the claimant's HMM. HMMs are more complicated to implement compared to GMM and provide better results in text-dependent systems, but the NIST Speaker Recognition Evaluation (SRE) results that are discussed in Section 7.8 have not shown any benefits of HMM over GMM in text-independent systems.

7.6.3 Other Methods

Neural networks and support vector machines (SVM) have been explored as a solution to the model classification problem. Neural networks are explicitly modeled to discriminate between the genuine user and a set of imposter users. Even though neural networks have an extremely flexible architecture, creating the training model requires a large amount of data from both the genuine and imposter datasets. An inaccurate or a small imposter dataset can lead to degradation in the performance of neural networks. SVM classifiers are becoming increasingly popular since they are capable of classifying user features in multidimensional space using nonlinear boundaries. SVM classifiers model a boundary between genuine users and imposters instead of separately modeling the probability distributions of the genuine users and the imposter population. Typically, the imposter population is held constant, similar to the UBM, which is provided as input to the SVM training function with the genuine user's speech samples. The output of the function maps to one of two classes: genuine or imposter. Hybrid techniques that combine SVM and GMM are being researched for improving speaker verification results. Vector quantization (VQ) algorithms have shown good results in speaker recognition [8–10]. Vector quantization algorithms build a codebook for each user based on the user samples. The basic premise of VQ algorithms is to create a codebook from a large number of spectral vectors using different training algorithms. The LBG algorithm has proven to be an effective codebook-generating algorithm used in the VQ algorithm for speaker verification [11].

7.7 Standards

At the time of this writing, the draft of ISO/IEC 19794-13 Voice Data was being progressed through the ISO/IEC SC 37 Committee. The proposal standardizes the data storage format of raw data as well as extracted features. This could include audio signals that are already encoded or extracted using different techniques. The proposed standard also allows the inclusion of metadata describing ambient conditions when the audio signal was captured.

INCITS M1 is also drafting a standard on specifying a data format for exchanging raw audio data. The goal of this standard is to specify the data format in XML along with all the attributes required for generating the template. The specification will be in XML and interoperable with applications based on VoiceXML 3.0. At the time of this writing, the INCITS M1 draft standard was not related to the ISO/IEC 19794-13 proposal.

Currently, there are no standards for speaker recognition that target data acquisition equipment such as microphones or restrictions on ambient noise.

7.8 Evaluations

The Speech Group at NIST has played a crucial role in assessing the current state of the art in text-independent detection and identifying challenge areas through its series of Speaker Recognition Evaluation (SRE). Since 1996 NIST has conducted several SREs with the latest one in 2010 [12]. Evaluating text-independent speaker recognition systems requires a speech dataset consisting of multiple users, collected over a period of time using telephone handsets of different makes and models to simulate a real-world scenario. The Linguistic Data Consortium (LDC), founded in 1992, collected a series of datasets called the Switchboard and MIXER datasets. The dataset was initially used in government sponsored research for speaker and topic detection, but was later opened up for research activities. These datasets have formed the basis of the SRE campaign. For a detailed explanation of these datasets, readers are encouraged to read [13].

The main focus of each SRE is on the task of speaker detection. Although the task has remained the same over the years, additional test cases were added to each evaluation. The initial SRE focused on one-speaker detection where the sample consists of speech from only one speaker in a conversation. Later on, two-speaker detection was introduced as a test case, where the sample consists of conversation between two speakers out of which only one speaker needs to be detected. The duration of the speech samples have also become a variable of interest along with different types of microphones and landline and cellular channels.

The performance of algorithms participating in each SRE is evaluated in terms of a cost function where the cost parameters of a missed detection and a false alarm are fixed by NIST. The NIST evaluations assign a higher cost to a missed detection compared to a false alarm. The cost function described in (7.2) has been used for evaluating algorithms participating in the SRE campaign. The probability of a missed detection ($P_{\text{Miss|Target}}$) and false alarm ($P_{\text{FalseAlarm|Nontarget}}$) are generated as part of the evaluation for each algorithm.

$$C_{\text{Det}} = C_{\text{Miss}} * P_{\text{Miss|Target}} + C_{\text{FalseAlarm}} * P_{\text{FalseAlarm|NonTarget}} * P_{\text{NonTarget}}$$
$$C_{\text{Miss}} = 10, C_{\text{FalseAlarm}} = 1, P_{\text{Target}} = 0.01, P_{\text{NonTarget}} = 0.99$$

(7.2)

Each SRE has shown a reduction in EER from the preceding evaluation even though the test conditions have increased in their level of difficulty. The SRE 2008 produced an EER of less than 2% for the state-of-the-art systems for the speaker detection tasks [14]. At the time of this writing, the SRE 2010 results had not yet been published. SRE 2010 is using the MIXER dataset, which contains speech data collected over landline, cell phones, and room microphones. Along with telephone conversations and interview speech recordings, the data was also categorized into high and low vocal efforts of the speaker. The evaluation is being conducted on nine different test scenarios that are testing channel effect, effort effect, and sample length effect on the participant's algorithms. The SRE 2010 also includes, for the first time, the Human Assisted Speaker Recognition Test (HASR), which will be performed by the systems involving human expertise to make the final decision about the trial. The level of human involvement is not specified, so systems could automate the bulk of the processing or perform minimal automated processing on the data. Forensic applications that require human intervention are the core audience for the results of HASR.

In 2003 the Netherlands Forensic Institute (NFI) and the Netherlands Organization for Applied Scientific Research (TNO) published a report that analyzed field data collected from wiretaps on individuals being monitored for criminal intent [15]. The main objective was to identify individuals using a one-speaker and two-speaker detection methodology. Fifteen systems were used to analyze the data, and the lowest EER was 12.1% and the highest EER was 35%. The results were also analyzed based on the duration of enrollment and test samples and one-speaker or two-speaker samples. The results showed that longer duration enrollment samples were more significant in improving performance than longer duration test samples. The negative impact of introducing a two-speaker sample was larger if used in test samples compared to the enrollment samples.

In 2004 IBG conducted a speaker verification test on three commercially available systems predominantly designed to address authentication issues for financial institutions [16]. A total of 219 subjects participated in this test, which was conducted over two sessions, although every subject did not complete all sessions. In the first visit the subjects were enrolled using a landline and then verified using a landline and a cell phone. In the second visit the subjects verified against their initial enrollment using a landline and a cell phone. The enrollment templates of 20 users were used in generating imposter match scores. Overall the systems worked significantly better over landlines compared to cell phones. The lowest FMR and FNMR, 0.586% and 0.800%, respectively, were for verification using landline during the first visit. In comparison, cell phone verification during the first visit produced an FMR and an FNMR of 5.863% and 4.902% [17].

In 2009 IBG released the *Comparative Biometric Testing Round 7 Public Report*, which described the performance evaluation of a commercially available voice recognition system, the AGNITIO Automated Speaker Identification System. Data of two durations were used for testing: 15 seconds and 60 seconds. These data were collected over telephone and microphone channels. Identification rates were computed for all possible scenarios against an enrolled database of 500 samples. The results are summarized in Table 7.2 [18].

Mobile Biometry (MOBIO) was initiated as part of the European Union (EU) Seventh Framework Research Programme with the aim of analyzing face and speaker recognition systems by creating a testbed of voice samples captured from mobile devices. The results of the first MOBIO evaluation were presented in 2010 [19]. The MOBIO database was collected using a mobile device at six different sites in five European countries and five different participants provided their systems

Table 7.2 Speaker Identification Rates from IBG CBT Round 7

Rank	Intra Telephone %		Intra Mic %		Mic Versus Telephone DB %		Telephone Versus Mic DB %	
	60 seconds	15 seconds	60 seconds	15 seconds	60 seconds	15 seconds	60 seconds	15 seconds
Rank 1	99.26	96.06	98.31	97.34	99.51	97.07	97.56	95.61
Rank 2	100	98.77	99.76	99.03	99.76	98.53	99.51	98.78
Rank 3	100	99.26	100	99.03	100	99.02	100	99.27

Source: [18].

for evaluation. The data were collected from six sites in five different countries in Europe; 160 participants from across these sites completed six separate data collection sessions. In each session the subjects were asked questions and their responses were categorized into three sets: (1) predetermined response, (2) reading responses from an answer sheet, and (3) free speech response. A performance metric called the half total error rate (HTER), which is the average of the FAR and the FRR, was used to assess the performance of the systems. The datasets were categorized based on the gender of the users, and the HTER results are shown in Table 7.3.

Table 7.4 lists the research databases that have been used extensively for speaker recognition evaluations.

7.9 Applications and Trends

Speaker recognition has found a willing market in telephony authentication for transaction security. Telebanking and telemedicine are examples of transactions that require user authentication and speaker recognition provides that functionality without the user having to buy additional hardware.

British Telecom (BT) started using speaker recognition as part of its URU online identity verification platform. The URU platform is meant to reduce identity fraud and can be used as part of a Web application. For example, a user will provide his or her username and password at a Web site. If the username and password are accepted, a phone call will be placed to the user's preferred phone and a speaker verification check will occur. If the verification passes, the user is allowed to access the Web site.

Home Shopping Network (HSN) deployed one of the earliest speaker verification systems in order to reduce fraud and inconvenience. HSN was already using

Table 7.3 MOBIO Speaker Verification Results

Algorithm	HTER %		
	Male	Female	Average
1	10.47	10.85	10.66
2	14.49	15.7	15.1
3	13.57	15.27	14.42
4	15.45	17.41	16.43
5	11.18	10	10.59

Source: [19].

Table 7.4 Speech Sample Research Databases

Name	Number of Subjects	Description
TIMIT	630	Read speech containing 10 phonetically diverse sentences recorded using a high-quality microphone
KING	51	Speech of 30 seconds on assigned topics recorded 10 times using narrowband and wideband telephone channels
YOHO	138	Consisted of four enrollment and 10 verification sessions; in each session user had to speak a sequence of three two-digit numbers
Switchboard 1	543	Total of 2,400 two-sided conversations, selected from a list of 55 topics and lasting at least 5 minutes
Switchboard 2	P1—657 P2—679 P3—640	Recording of telephone conversations between individuals from the same region of the United States, and mainly college and postcollege demographics. Multiple types of landline instruments used
Switchboard Cellular	P1—254 P2—419	Recording of telephone conversations of individuals speaking on cell phones
MIXER—1 and 2	600 speakers with 10 or more calls	Recording of telephone conversations; speakers included nonnative English speakers and 200 speakers recorded conversations using four different types of channels
MIXER—3	1,867	Recording of telephone conversations of 15 or more calls, included speech samples in 19 different languages
MIXER—4	200	Recording of telephone conversations of 10 calls; primary language of use was English, but also included different languages
MIXER—5	300	Conversation in structured interviews of 30 minutes and reading from provided text; speech recorded using 12 microphones
XM2VTS	295	Speech data recorded over 4 months
BioSecure Multimodal	More than 600	Recorded speech samples in office environment and outdoor environment using mobile devices and over the Internet; two acquisition sessions separated by periods of 1 to 3 months
Gandlaf	86	24 Swedish language telephone conversations per subject; used multiple types of landline and cell phone handsets
MOBIO	160	Recorded speech samples using mobile devices; dictated answers, predetermined responses, and conversational speech samples

Source: [13].

speech recognition to allow customers to place orders over the phone. In 1999, HSN started using speaker recognition to retrieve customers' profile information and authenticate them. The customers would have to speak their phone numbers and if they were already enrolled, verification would be conducted. If they were not enrolled, they would be guided through the enrollment procedure and use it for future purchases.

The New York City Department of Probation started using speaker recognition in 1999 as part of their probationers' monitoring system. Probationers enrolled in the program had to call an automated system from a predetermined phone number and location and perform the verification process.

Speaker recognition can be conducted covertly without the knowledge of the user. This has led to the use of speaker recognition in intelligence and investigation agencies in identifying people of interest. These applications operate in an identification mode and have to use text-independent speaker recognition or segment the voice signal to identify overlaps with a previously recorded voice of a known individual.

The combination of speech recognition and speaker recognition offers a multifactor authentication based on knowledge and biometric characteristics. For example, an authentication system initiates the process by asking a user a series of questions. Each answer provided by the user is interpreted by the speech recognition component and checked against the answers provided during enrollment. Speaker verification is also conducted in parallel on the voice signal. The combination of decisions from the two systems is then used to make a final decision. This approach can overcome operational limitations of speaker recognition and increase overall security.

7.10 Design and Deployment Considerations

Voice recognition has several user and environmental challenges to consider. The ease of data capture and the distributed implementation of speaker recognition also contribute to its operational challenges, which are discussed in this section.

7.10.1 Voice Variations

A person's voice characteristics are affected by a multitude of factors such as anxiety, health, stress, speech volume, outside temperature, and speech rate. All of these have an impact on the quality and consistency of the voice sample. This variability is a particularly important issue in geographical regions that experience frequent climate change and thus make users more susceptible to health issues impacting a person's voice. Research conducted on emotional voice variability in speaker verification has shown a strong correlation in increase of equal error rates and acoustic parameters affected by 10 different speaking styles [20]. The same research also concluded that including voice samples representative of stressful conditions in the training model improves the overall recognition rate.

There is an impact of time on human voice. The longer the time between enrollment and verification, the higher the probability of a matching error [21]. Although the change might be minimal, adaptation models are required that account for the changes over time in a person's voice and minimize matching errors.

7.10.2 Background Noise

Noise generated in the background distorts the voice signal and adds extraneous information that can suppress the features of interest. Background noise is generated by a variety of sources: traffic in the background, voices of other people, radio sound, and office noise from the ventilation system. Although it is difficult to restrict users to a noise-free environment, implementers must take special care to compensate for background noise. Stationary noise, which is repetitive, can be filtered by using spectrum subtraction techniques. Nonstationary noise is more difficult to filter and can interfere with the feature extraction process. Room acoustics can create echoes and reverberations and affect the voice signal captured by the microphone. Reverberation is observed in situations where the microphone is placed away from the speaker, which can degrade recognition performance as well [22].

7.10.3 Channel Effect

Until a decade ago channel effects were a nonissue because of the prevalence of landlines in telecommunication infrastructure. Mobile phone and voice over IP (VoIP) usage has increased tremendously and will become the primary means of communication in the near future. Audio signals transmitted over landlines, mobile phone networks, and VoIP systems are encoded differently, which affects the spatial relation of cepstral features. A person who enrolls on a landline and then attempts to verify on VoIP channel has a higher probability of generating a matching error [23]. A solution recommended by implementers is to create different enrollment templates for each type of communication channel and use them accordingly for verification.

7.10.4 Microphone Effect

All microphones essentially do the same thing—convert varying pressure waves in the air into electrical signals. Every microphone has its own transfer function that changes the spatial relation of the cepstral coefficients. Microphones can be typically categorized into five different groups based on how the electrical signals are produced and measured, and the cepstral coefficients for the same person can vary from one microphone to the other. NIST used multiple types of microphones as part of SRE 2006 that included the following types: earbud/lapel, mini-boom, courtroom, conference room, distant, near-field, PC stand, and microcassette. Various techniques have been explored that normalize voice signals from different microphones, but this still remains an implementation issue.

7.10.5 Duration of Samples

The length of the voice sample and the number of repeat samples have an impact on verification performance. Research conducted by NIST as part of the SRE campaign has shown that longer duration samples in enrollment and verification results in improved performance for all types of speaker recognition systems [24]. In the same research it was observed that speech durations longer than 15 seconds did not result in performance gain, but speech duration had an impact on samples less than 15 seconds. Empirical data suggests that the enrollment procedure should include a number of repetitions and voice samples longer than at least 3 seconds for robust performance.

7.10.6 Recording Attack

For speaker recognition a playback recording attack is the simplest one to conduct due to the availability of sound recorders. However, the ease of attack is not correlated to the success rate of such an attack. Most sound recorders cannot capture all the frequency information of the human voice and thus are usually not successful. Voice samples, just like any other biometric trait, show variability from one sample to the next. Playback attack detectors are based on the premise that if two utterances are too similar, then an attack has occurred. Although this is a difficult system to implement due to the requirements of storing every single interaction, research is being conducted to create such operational systems. Deploying a text-prompted system that is capable of performing a challenge response is also an effective countermeasure.

7.11 Summary

Speaker recognition has huge potential for user authentication in the telecommunication context. Researchers are starting to look at new areas of applicability and improvement of this technology. Higher levels of speech information have not been used extensively in speaker recognition and preliminary exploration has shown that it can improve accuracy. NIST's SRE campaign has been extremely influential in driving the progress of this technology and new benchmark datasets are required to focus on real-world robustness. Text-independent systems are of great interest to law enforcement and surveillance communities, but are unlikely to be used in consumer-facing applications in the near future. VoIP applications such as Skype encrypt their audio communication, which poses a challenge for speaker recognition. Speech recognition applications are being increasingly deployed, which will also spur the adoption of speaker recognition, specifically in financial institutions and healthcare applications. The ubiquity of mobile devices will likely be the biggest driver of speaker recognition in the near future.

References

[1] Fletcher, H., "The Nature of Speech and Its Interpretations," *Bell Syst. Tech. J.*, Vol. 1, 1922, pp. 129–144.

[2] Kersta, L., "Voice Print Identification," *Nature*, Vol. 196, 1962, pp. 1253–1257.

[3] Bogert, B. P., M. J. R. Healy, and J. W. Tukey, "The Frequency Analysis of Time Series for Echoes: Cepstrum, Pseudo-Autocovariance, Cross-Cepstrum, and Safe Cracking," *Time Series Analysis*, 1969, pp. 209–243.

[4] Doddington, G., "Speaker Recognition Based on Ideolectal Differences Between Speakers," *EUROSPEECH-2001*, 2001, pp. 2521–2524.

[5] Bimbot, F., et al., "A Tutorial on Text-Independent Speaker Verification," *EURASIP Journal on Applied Signal Processing*, Vol. 2004, 2004, pp. 430–451.

[6] Oppenheim, A. V., and R. W. Schafer, *Discrete Time Signal Processing*, Englewood Cliffs, NJ: Prentice-Hall, 1989.

[7] Campbell, J., "Speaker Recognition," in *Biometrics: Personal Identification in Networked Society*, A. K. Jain, R. Bolle, and S. Pankanti, (eds.), New York: Springer, 1999.

[8] Reynolds, D. A., "A Gaussian Mixture Modeling Approach to Text-Independent Speaker Identification," Ph.D. dissertation, Georgia Intsitute of Technology, 1992.

[9] Soong, F., et al., "A Vector Quantization Approach to Speaker Recognition," *AT&T Technical Journal*, Vol. 66, 1987, pp. 14–26.

[10] Rosenberg, A. E., and F. K. Soong, "Evaluation of a Vector Quantization Talker Recognition System in Text Independent and Text Dependent Models," *Computer Speech and Language*, Vol. 22, 1987.

[11] Linde, Y., A. Buzo, and R. M. Gray, "An Algorithm for Vector Quantizer Design," *IEEE Transactions on Communications*, Vol. 20, 1980, pp. 84–95.

[12] NIST, "Speaker Recognition Evaluation," http://www.itl.nist.gov/iodmig/tests/sre.

[13] Martin, A. F., "Speaker Databases and Evaluation," in *Encyclopedia of Biometrics*, S. Lee and A. K. Jain, (eds.), New York: Springer, 2009, p. 14.

[14] Reynolds, D. A., and W. A. Campbell, "Text-Independent Speaker Recognition," in *Handbook of Speech Processing*, J. Benesty, M. M. Sondhi, and Y. Huang, (eds.), New York: Springer, 2008, p. 779.

[15] Vanleeuwen, D., et al., "NIST and NFI-TNO Evaluations of Automatic Speaker Recognition," *Computer Speech & Language*, Vol. 20, April 2006, pp. 128–158.

[16] IBG, *State of Speaker Verification Technology Report*, New York, 2004.

[17] Nanavati, S., "Biometric Testing," *RSA Conference 2006*, San Francisco, CA, 2006.

[18] IBG, *Comparative Biometric Testing Round 7 Public Report*, New York, 2009.

[19] Marcel, S., et al., *On the Results of the First Mobile Biometry (MOBIO) Face and Speaker Verification Evaluation*, IDIAP Research Institute, 2010.

[20] Klasmeyer, G., et al., "Emotional Voice Variability in Speaker Verification," *ITRW on Speech and Emotion*, Newcastle, U.K., 2000.

[21] Wang, L., and T. F. Zheng, "Creation of Time-Varying Voiceprint Database," *Oriental COCOSDA 2010*, Nepal, 2010.

[22] Akula, A., and P.L.D. Leon, "Effects of Room Reverberation on Speaker Identification," *EUSIPCO 2008*, 2008.

[23] Elliott, S., and A. Rolfe, "Case Study Phone Based Voice Biometrics for Remote Authentication," *RSA Conference 2007*, San Francisco, CA, 2007, p. 41.

[24] Przybocki, M. A., and A. F. Martin, "The 1999 NIST Speaker Recognition Evaluation Using Summed Two-Channel Telephone Data for Speaker Detection and Speaker Tracking," *Digital Signal Processing*, Vol. 10, 2000, pp. 1–18.

CHAPTER 8
Vascular Pattern Recognition

Vascular pattern recognition is based on the unique structure formed by the network of blood vessels. The structure of blood vessels is formed during embryonic stage and remains relatively constant throughout the lifespan of the individual. Currently all commercially available vascular pattern recognition technologies are based on vein patterns, and the two tend to be used interchangeably. This technology uses the difference in the absorption properties of oxygenated and deoxygenated hemoglobin in human blood under infrared illumination. Deoxygenated hemoglobin has a higher absorption index compared to oxygenated blood and the tissues surrounding the blood vessels. When infrared light is incident on human skin, the deoxygenated hemoglobin in veins absorb more of the infrared light than the surrounding region and the oxygenated blood in the arteries. In the image of this illuminated region captured by a camera sensitive to infrared wavelength, the veins appear darker and forms a 2-D vascular pattern, as shown in Figure 8.1. This technology has several benefits:

1. Veins are present all over the body, which allows data capture from multiple body parts.
2. Veins are located underneath the skin and are not affected by abrasions to the outer skin layer.
3. The imaging technique does not require physical interaction with the body part and thus has fewer hygiene concerns.
4. The flow of blood can be used as a liveness detection measure.

Currently there are three different parts of the hand that are used in commercially available vein recognition vendors: the finger, the back of the hand, and the palm of the hand. There is ongoing research evaluating the feasibility of utilizing facial vascular patterns, comprised of arteries, which is discussed in Section 8.4. Vein recognition technology is a relatively new technology that is gaining traction in consumer-facing applications in financial and healthcare sectors, especially in the East Asia region. This chapter discusses the underlying principles of vascular pattern recognition, deployment challenges, and the future direction of this technology.

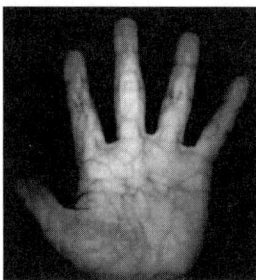

Figure 8.1 Infrared image of vascular pattern.

8.1 History

The first recorded suggestion of the use of vein patterns for person recognition can be traced to a patent application filed by Joseph Rice in 1985 [1]. While working for Eastman Kodak, Mr. Rice observed the capability of capturing vein patterns using infrared illumination and its distinctiveness among individuals. His patent application "Apparatus for the Identification of Individuals" was approved in 1987, although commercial products based on this discovery only appeared in 2000. In 1992 Dr. K. Shimizu discussed the possibility of capturing vascular patterns with computerized tomography (CT)–based optical transbody imaging [2]. A research paper published in 2000 described the feasibility of vascular pattern recognition as a biometric technology [3]. This research captured the subcutaneous vein patterns from the back of the hand and was also commercialized in 2000. A patent was awarded in 2001 that described in detail the use of the back of the hand vascular pattern for personal identification [4]. Vein pattern recognition is an active area of research, and a project to create an international standard ISO/IEC 19794-9 for data capture and exchange was published in 2007, thus indicating the growing relevance and acceptance of this technology [5].

8.2 Vein Pattern Acquisition

Vein pattern imaging sensors typically use infrared illumination of wavelength in the range of 700–1,200 nanometers (nm) since it can penetrate human tissue up to 3 millimeters [6]. Infrared illumination of the wavelength outside of this range does not provide enough penetration and clarity of vein patterns in the resulting image. There are two main types of vein pattern imaging: *reflective* and *transmissive*. In reflective imaging systems, shown in Figure 8.2, the illumination source and imaging sensor are placed on the same side of the target area. The illumination source projects infrared light onto the skin surface and the imaging sensor captures the reflected light from the subcutaneous region. Current commercial technologies use reflective imaging for capturing vein patterns from the back of the hand and the palm of the hand. In the transmissive imaging systems, shown in Figure 8.3, the illumination source and imaging sensor are placed on opposite sides of the target area. The illumination source projects light onto the skin surface and the imaging sensor captures the light that passes through the tissue. It is difficult to capture vein pat-

8.3 Feature Extraction

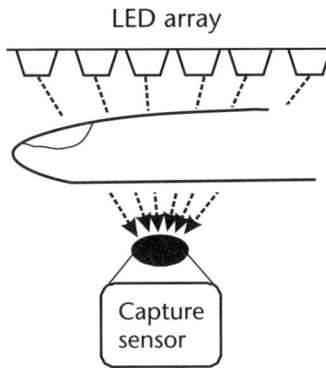

Figure 8.2 Transmissive imaging.

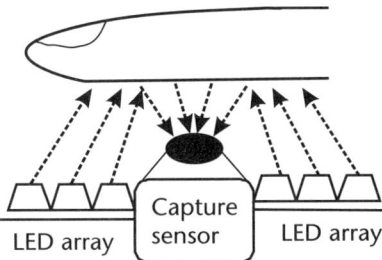

Figure 8.3 Reflective imaging.

terns when infrared light passes through thicker parts of the body, and this method is used on regions that are smaller and thinner, such as the finger.

Infrared light emitting diodes (LED) are commonly used for illumination. LEDs have low power consumption, are inexpensive to manufacture, and are small enough that they can be arranged in a variety of shapes to ensure the proper capture of vein patterns. As reflective imaging systems combine the illumination source and the imaging sensor into the same device, they can be comparatively smaller than transmissive imaging systems.

8.3 Feature Extraction

In the resulting image captured by the sensor, the vein patterns appear darker than the surrounding region. Using image processing techniques, the veins patterns are enhanced so that they are clearly differentiated from the surrounding region. The resulting image is analogous to a fingerprint image, where the vein structure is similar to the dark ridgelines, and the surrounding tissue region is similar to the valleys.

The feature extraction process can be divided into the following steps: region of interest (ROI) extraction, noise reduction, and segmentation. The ROI extraction process separates the background of the image from the ROI, which, depending on the area of the hand, can be either the finger, the back of the hand, or the palm. The reflective properties of human tissue and the imaging mechanism introduce a certain level of noise that has to be removed to ensure a high-quality image.

Gaussian blur filters are a commonly used technique for smoothing images, but they tend to reduce the sharpness of the edges in the image. Research has shown that edge-preserving filters based on nonlinear diffusion are an effective means of maintaining the fidelity of the vascular pattern [7]. Extraction of vein edges is the final step and the type of extraction is dependent on the matching methodology.

8.4 Feature Matching

Feature-matching techniques can be loosely categorized into two groups: edge matching and point matching. In edge-matching techniques a degree of similarity is calculated based on the correlation of the vein edge pattern of the two images. A ratio of the overlap of pixels between two vein patterns is calculated to determine the degree of similarity. The vein pattern can also be viewed as a 2-D graph and graph-matching techniques are used for comparing the two vein patterns. Point matching is similar to minutiae matching in fingerprints [8]. The vein network, similar to fingerprint ridgelines, is formed from lines that end and bifurcate abruptly. These discontinuities in the vein network are described by their x-coordinate and y-coordinate in 2-D space along with the type of discontinuity. The spatial relationship between these points, along with other attributes, is used to generate a similarity score.

8.5 Facial Vascular Patterns

Ongoing research is evaluating the use of vascular patterns in the face region for recognition. It should be noted that since this is an extremely new area, scientific publications may also list this technology under face recognition. This method is based on the principle of heat diffusion between blood vessels and the surrounding tissue area. Using thermal imagery, the superficial blood vessels can be extracted by analyzing the temperature values around them. Thermal imaging of vascular patterns requires using parts of the body that have a high concentration of blood vessels adjacent to a low concentration of blood vessels. Research has shown that the face provides such a region, and the experimental results on a small dataset were promising [9]. Although there are operational challenges such as the cost of deployment, usability, and environmental effects that need to be studied, facial vascular pattern recognition presents another avenue of growth for vascular pattern recognition.

8.6 Commercially Available Technologies

Currently, commercially available technologies use three different parts of the hand for extracting vein patterns: the finger, the palm of the hand and the back of the hand. Hitachi Ltd., based in Japan, introduced technology based on the finger vein in 2004. This system uses transmissive imaging and thus requires an enclosure to prevent ambient lighting from interfering with the capture sensor, as shown in

8.7 Standards

Figure 8.4 Hitachi Vein Reader system. (*Source:* Hitachi, Ltd.)

Figure 8.4. This technology is used both for logical and physical access control applications.

Fujitsu Ltd., based in Japan, introduced technology that uses reflective imaging to capture vein patterns from the ventral side of the hand. Fujitsu has integrated palm vein recognition technology into a computer mouse that is used for user recognition. Since this technology uses reflective imaging, the entire system is small enough that it can be attached as a peripheral device or integrated into small-scale devices. This device is shown in Figure 8.5.

Techsphere Ltd., a company based in the Republic of Korea, introduced back-of-the-hand vein recognition technology in 2001. This device captures the vein pattern from the dorsal area just below the knuckles of the hand using reflective imaging. The entire system has a relatively larger form factor than the other two commercial technologies and is currently deployed only for physical access control applications (see Figure 8.6).

8.7 Standards

The ISO/IEC 19794-9:2007, released in 2007, defines the data exchange format for human vascular pattern image data. The main objective of this standard is to specify a set of requirements for formatting raw or processed vascular pattern images so that applications can interpret data generated by a different vendor. Since the standard is defined to support commercially available technologies, it currently only specifies the exchange of finger, palm, and back-of-the-hand vein pattern data. The standard does have the capability of accommodating vascular pattern images

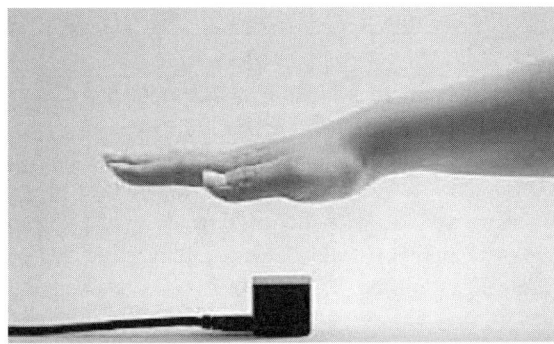

Figure 8.5 Fujitsu PalmSecure™ system. (*Source:* Fujitsu Limited.)

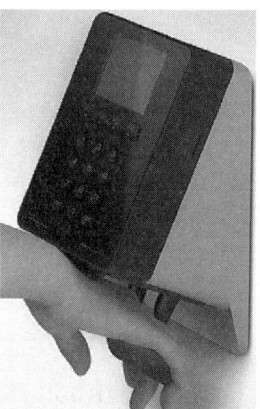

Figure 8.6 Techsphere VP-II X system. (*Source:* Techsphere. Reprinted with permission.)

from other parts of the body. Table 8.1 summarizes some of the data exchange requirements specified in ISO/IEC 19794-9 [5].

8.8 Performance Evaluations

Fujitsu Ltd. has published data from internal tests conducted on palm vein recognition technology. Approximately 140,000 palm vein images were collected from 70,000 individuals. All users provided three samples for generating the enrollment template and one sample for verification. In their analysis FAR of 0.00008% and FRR of 0.01% was reported [10]. Other details of this test, such as demographic information and matching threshold, are not publicly available.

In 2006 the International Biometric Group (IBG) conducted a comparative test of multiple biometric systems that includes the Hitachi UBReader TS-E3F1 system and Fujitsu PalmSecure™ system. The results are summarized in a report titled *Comparative Biometric Testing Round 6 Public Report* [11]. Data was collected from 650 individuals over two different sessions, and the final analysis was conducted on approximately 20,000 samples. Due to the inherent differences in the decision-making mechanisms of the two systems, a direct comparison of the results cannot be made and readers who read the report should be cognizant of

Table 8.1 Summary of ISO/IEC 19794-9:2007

Field	Specification
Spatial resolution	No minimum specified
Grayscale depth	7-bit dynamic grayscale depth for images
Illumination types	Illumination type must be specified and is not limited to any certain type
Imaging type	Reflective or transmissive type must be specified
Image compression	JPEG, lossless JPEG, JPEG2K, and raw
Body areas	Dorsal and ventral side of all fingers on the hand, dorsal side of the hand, palm of the hand

Source: [5].

this difference. The Hitachi system provided similarity scores for each comparison, whereas the Fujitsu system provided a decision at three different threshold levels. An FMR of 1 in 10,000 was observed in both systems, and true accept rates of 97.23% and 99% were observed in the Hitachi and Fujitsu systems, respectively. Another aspect of the test analyzed the recognition rates for samples collected on the same day as enrollment and for samples collected on different days. A significant difference in FMR and FNMR was observed in the same day and different day verification attempts for both systems. This increase in error rates was attributed to a change in user interaction and habituation effect due to the difference in the time of enrollment and the verification attempts.

8.9 Applications and Trends

The early adopters of this technology focused on deploying it in consumer applications for logical and physical access. The compact form factor of palm and finger vein recognition devices makes it a good fit for password and token replacement and password management applications. Japan's Bank of Tokyo-Mitsubishi first introduced finger vein recognition in their ATMs in October 2004. Several other major financial institutions have also introduced this technology, and a survey indicated that approximately 80% of institutions that use biometrics use some form of vein recognition technology [6].

The digitization of medical records has brought focus on the importance of accurate patient identification. Vein recognition technologies have found traction in this area—Carolinas Healthcare System (CHS) is an example of one such deployment. CHS is using palm vein recognition to handle patient check-in, electronic healthcare records, and other patient-related processes such as health insurance audits. The e-healthcare initiatives in the United States are expected to be one of the biggest drivers for this technology.

Several real-world deployments are using vein recognition for time and attendance and employee management applications. The seaport located in Halifax, Canada, is one of the largest seaports in North America with more than 4,000 employees accessing the premises through over 2,000 access points. Controlling physical access in challenging weather conditions was a key factor in deciding which technology to use, and eventually back-of-the-hand vein recognition technology was selected for deployment [12]. Traditional employee time and attendance systems use cards for managing employee attendance. This technology is being replaced with vein recognition technology as it provides a higher level of auditability and nonrepudiation.

Some novel applications of this technology have exhibited future potential of this technology. One such demonstration at the 2005 Tokyo Motor Show used finger vein recognition for authenticating the driver who started the car and configured the seat height and steering wheel height according to the personalized settings of the user. Such applications are likely to proliferate in the upcoming decade.

8.10 Design and Deployment Considerations

Experiments and lessons from operational deployments have highlighted factors that have an impact on performance of vein pattern recognition. Environmental conditions such as excess ambient light can interfere with the imaging system and result in a blurred image. This issue is typically addressed by building an enclosure around the illumination source and the imaging system.

Physical features such as hair can interfere with the resulting image and affect image processing and subsequent feature extraction. The finger area and the palm of the hand are not affected by this, but the back-of-the-hand systems have to address this issue with a subset of the user population. Certain types of tattoo inks also absorb infrared illumination and prevent a proper image from being captured [6].

Occupations that result in hands getting covered in carbon-based substances such as soot or charcoal will also have a negative impact on image capture. Carbon-based substances absorb infrared illumination and interfere with the imaging process.

All the commercial technologies described in this chapter require images from the same region of the finger, the palm, or the back of the hand for successful recognition. Although no physical contact is required between the body part and the acquisition sensor, the system needs to capture the same region of interest consistently. For example, if a finger vein system captures the vein pattern between the first and second segment from the index finger, the same region has to be presented at every successive capture. A different region from the same finger will result in a nonmatch. Although human interaction is not a significant limitation to commercial adoption, the requirement of capturing the same region of interest without any physical contact makes it difficult to find a solution that works for the entire population. In addition, infrared imaging increases the difficulty for a user or supervisor to ascertain if the correct region of interest is being presented. Technologies such as face recognition provide an unambiguous view to the user or the supervisor and the appropriate presentation can be made based on the visual feedback. Currently, human interaction is resolved using a guidance mechanism such as a cradle that aids the user in presenting his or her hand in a consistent manner. Ongoing research in improving usability of biometric systems is attempting to make this a truly noncontact technology.

8.11 Summary

The last decade has seen major advances in the field of vascular pattern recognition and an increasing number of deployments in consumer-facing applications. With increasing use, new operational challenges will come to light, but ongoing applied research interest in this technology will eventually solve these challenges. The overlap of the same region with other biometric modalities such as fingerprints, palm prints, and face also indicates multimodal fusion as another growth driver for this technology. Although there is evidence from the medical field that vein patterns remain permanent over the life of an individual, this has not been empirically tested through any longitudinal studies. The financial and healthcare sectors have been the major drivers for this technology until now and will likely remain so in the near future.

References

[1] Rice, J., "Apparatus for the Identification of Individuals," U.S. Patent No. 4699149, 1987.

[2] Shimizu, K., "Optical Trans-Body Imaging—Feasibility of Optical, CT and Functional Imaging of Living Body," *Medicina Philosophica*, Vol. 11, 1992, pp. 620–629.

[3] Im, S. -K., et al., "Biometric Identification System by Extracting Hand Vein Patterns," *Journal of the Korean Physical Society*, Vol. 38, 2000, pp. 268–272.

[4] Choi, H. -S., "Apparatus and Method for Identifying Individuals Through Their Subcutaneous Vein Patterns and Integrated System Using Said Apparatus and Method," U.S. Patent No. 6301375, 2001.

[5] ISO/IEC, *ISO/IEC 19794-9 Information Technology—Biometric Data Interchange Formats—Part 9: Vascular Image Data*, Geneva, Switzerland, 2007.

[6] Wilson, C., *Vein Pattern Recognition: A Privacy-Enhancing Biometric*, Boca Raton, FL: CRC Press, 2010.

[7] Badawi, A. M., "Hand Vein Biometric Verification Prototype: A Testing Performance and Patterns Similarity," *IPCV*, 2006, pp. 3–9.

[8] Wang, Y., T. Liu, and J. Jiang, "A Multi-Resolution Wavelet Algorithm for Hand Vein Pattern Recognition," *Chinese Optics Letter*, Vol. 6, 2008, pp. 657–660.

[9] Buddharaju, P., I. Pavlidis, and C. Manohar, "Face Recognition Beyond the Visible Spectrum," Ch. 9, *Advances in Biometrics*, N. K. Ratha and V. Govindaraju, (eds.), New York: Springer, 2008, pp. 157–180.

[10] Watanabe, M., "Palm Vein Authentication," Ch. 5, *Advances in Biometrics*, N. K. Ratha and V. Govindaraju, (eds.), New York: Springer, 2008, pp. 75–88.

[11] IBG, *Comparative Biometric Testing Round 6 Public Report*, New York, 2006.

[12] Choi, A. H., "Back-of-Hand Vascular Recognition," in *Encyclopedia of Biometrics*, New York: Springer, 2009, pp. 55–60.

CHAPTER 9

Dynamic Signature Verification

A signature is commonly understood to be a mark or a sign made by an individual's own handwriting to signify approval, acceptance, contractual obligation, or proof of identity. The use of signatures in legal binding documents is a practice that has been going on for a very long time and has found acceptance in every facet of our lives. The Uniform Commercial Code (UCC), which tries to harmonize the various U.S. state government laws regarding commercial transactions, requires a written signature to enforce a contract exceeding $500. Traditionally, humans who perform signature verification use visual analysis of signature patterns and subjective judgment to ascertain if these signature patterns are from the same individual or not. Dynamic signature verification (DSV) uses information that captures the specific behavioral characteristics of a person when he or she is signing. The action of signing can be decomposed along a time line into multiple pen strokes, interspersed with extremely short breaks or a sudden change in direction of the stroke. This action is specific to an individual's natural action and acquired mannerisms. The human signing action is a type of ballistic motion [1]. In a ballistic motion the action performed by a set of human muscles becomes consistent with practice, such as a golf swing. This lends consistency to the signing action and uniqueness as well because of the human action. Research has shown that pen accelerations are consistent with force exerted by muscles in habitual signature [2]. DSV utilizes data such as the direction of strokes, the number of strokes, the velocity of different strokes, the overall shape, pressure, and other dynamically generated characteristics by the signature action.

DSV is a natural fit for applications which require strong biometric authentication. User acceptance is extremely high and training costs are low since the signature action is quite commonplace. DSV can easily integrate into existing applications without any retraining required for end users. In spite of its advantages, real-world deployments of this technology have been limited until now, but that is expected to change. According to a market report from Frost & Sullivan, revenues from the DSV market are expected to increase from $14.4 million in 2006 to $85 million in 2013 [3]. This chapter will discuss history, algorithm details, application trends, and challenges related to DSV technology.

9.1 History

The history of signature verification, in a forensic context, dates as far back as fingerprint recognition. In 1929 the first document describing forensic examination techniques and best practices for signature verification was published [4]. As technology advanced, automated methods of signature verification were developed, although these still focused on what the signature looks like and not on the dynamic characteristics exhibited by the signer. Readers interested in static signature analysis techniques should read the survey article published by Plamondon and Lorette [5]. With the advent of touch-screen technologies and faster computer processors in the 1970s, methods that were focused on the dynamic characteristics of signatures were developed. In 1971 a patent titled "Personal Identification Method and Apparatus" was awarded for positively determining the identity of an individual based on the electrical waveform generated when the person signs his or her signature [6]. In 1985 a patent was awarded for verifying an individual based on first- and second-order time derivates of pressure applied by an individual while signing [7]. Several commercial DSV products are currently available, and new research is constantly exploring methods of using touch-sensitive screens in PDAs, tablets, and smartphones for DSV.

9.2 Types of Signature Verification Systems

Signature verification systems are categorized based on the type of information being captured and processed by the system: off-line and online. In off-line signature verification a static signature (i.e., the graphical representation) is used for comparison. In online signature verification the dynamic features of the signature such as the position of the pen, stroke information, and other details are used for comparison. Online signature verification is used interchangeably with DSV.

9.3 Data Acquisition

Signature capture devices work on the principle of capturing data generated by the contact of a pen with a sensing surface. The most common method of capturing dynamic signature data is through the use of digitizer pads and a stylus. Such devices are designed to capture the x- and y-coordinates of the stylus as it moves over a digitizer pad, and more advanced devices can also capture the pressure and angle of tilt of the stylus. Figure 9.1 illustrates the time series representation of DSV characteristics. The resolution of the data, or the sampling rate, differs among the various digitizer pads available in the commercial market, which are summarized in Table 9.1. Some digitizers provide real-time graphical feedback on the screen as a person signs, whereas others wait for the signature to be completed before rendering it on the screen. Figure 9.2 shows examples of commercially available digitizer pads. With the increasing use of touch screens in computing devices, dynamic information can be captured from a variety of surfaces. Certain DSV systems use specialized pens that capture information such as the angle of the pen in addition to other dynamic information. Such devices have the look and feel of a regular pen

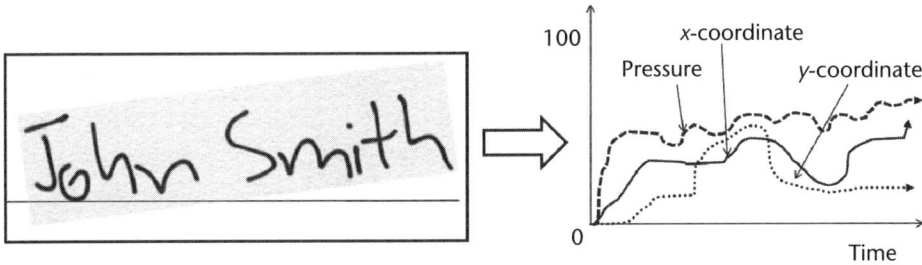

Figure 9.1 Dynamic signature.

Table 9.1 Summary of Digitizer Pads

Type	Description
Passive	Power generated by the digitizer pad
Active	Power generated by the stylus, which is used to transmit the signal to the digitizer pad
Optical	Camera integrated into the stylus that is used to capture the resulting signature
Electromagnetic	Electromagnetic signal produced by the stylus and measured by the digitizer pad; typically passive devices
Capacitive	Designed to use the difference in the electrostatic potential generated by the contact of the stylus with the digitizer pad

Figure 9.2 Digitizer pads. (*Source:* Wacom Technology Corporation, LLC. Reprinted with permission.)

with an onboard microchip that captures the dynamic characteristics. Although the sampling rates of these devices differ, most are capable of capturing more than 100 samples per second. The data acquisition technology is considered to be quite mature, which is reflected by the availability of several commercial products.

9.4 Feature Representation

DSV uses primary features such as time series measurements of x-coordinate, y-coordinate, and pressure applied by the tip of the pen on the surface, the number of strokes, the angle of the pen, and secondary features such as velocity and the acceleration of pen strokes. Some commercial products also incorporate the geometric shape of the signature in the pattern recognition algorithm as an enhancement. DSV feature representation is categorized into two general groups: *feature based* and *function based* [8]. Feature-based methods, also called global features, represent the signature as a vector of measurements such as average speed, acceleration, time duration, and angles. Several feature-based systems treat each pen stroke as a parameter that is described as a vector consisting of the following: starting x-coordinate and y-coordinate, ending x-coordinate and y-coordinate, the distance between successive strokes, the pressure at the beginning of the stroke, the pressure applied at the end of the stroke, and so forth. Function-based methods represent the signature as a time function of x-coordinates, y-coordinates, pressure, angle, acceleration, and velocity of the signature strokes. Higher-order derivates can be derived from the time functions and used as additional discriminatory information.

The characteristics described here represent a small collection of available characteristics; researchers have used upwards of 50 features in matching algorithms [9–11].

9.5 Feature Matching

Several different methods for matching two signatures have been examined in literature that can generally be categorized into *distance-based approaches* and *model-based approaches* [12].

Several distance-based approaches have been researched in DSV. Dynamic time warping (DTW) methods have proven to be successful at matching signatures. DTW algorithms are used for measuring similarity between two feature vectors that vary over time. Signatures have variable length even if they originate from the same user, which affects the time-based components of DSV. DTW minimizes a global cost function whose main objective is to normalize two signatures by identifying corresponding points and in the process and computes the difference in time alignment between the two signatures. The process can be interpreted as plotting a monotonic curve on a 2-D graph where one axis represents points from one signature and other axis represent points from the other signature. The global cost function signifies the degree of similarity between corresponding portions of the signatures. DTW can be explained using a simple example. In Figure 9.3 the x-axis represents the vector of measurements of the enrolled template and the y-axis represents the vector of measurements of the input sample that increase with time. The first and second elements of the two vectors match. The third and fourth elements of the enrolled template also match the second element of the input sample. This indicates a misalignment in the two sequences where the input sample is compressed in the time domain compared to the enrolled template. The fourth element of the enrolled template matches the third element of the input template. This indicates a compression in the input sample compared to the enrolled template. All the vector elements can be mapped onto the grid and a cost metric indicating the difference is calculated.

Euclidean and Mahalanobis distance classifiers have also been used to compute similarity scores between the reference and input signatures. Feature-based signature templates are predominantly used in this approach.

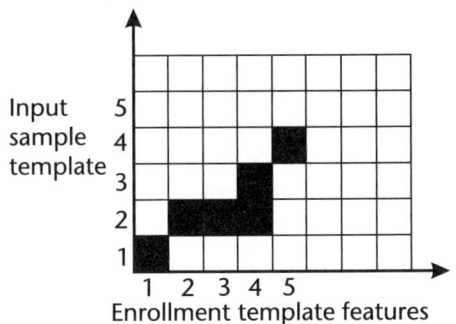

Figure 9.3 Dynamic time warping explanation.

Model-based approaches such as hidden Markov models (HMM) have proven to be successful in DSV. HMMs lend themselves well to signature verification because the signature features can be segmented into states and the time series provides state transitions. Matching methods that use discrete and continuous HMMs with Gaussian probability distributions are prevalent in research literature. Table 9.2 lists a subset of HMM-based DSV studies.

Statistical classifiers such as neural networks have been used in signature verification where data from genuine and imposter classes is used for training the network. Once the network is trained, it is used for classifying input data. The main drawback of neural networks is the requirement of the large amount of training data, which might be impractical in a real-world solution. Matching algorithms based on Gaussian mixture models (GMM) use a *universal background model* (UBM), which represents an average user and is adapted based on genuine user inputs to create a user model. GMMs are essentially a single-state HMM. GMMs with UBM have been used in speaker verification as well.

9.6 Standards

ANSI and ISO/IEC have published DSV standards at the U.S. national and international levels, respectively, to facilitate the interchange of signature data between different systems. At the U.S. national level, the ANSI/INCITS 395:2005 standard, published in 2005, specifies the data interchange format for the representation of the digitized sign or signature data. The data interchange format is generic and no application-specific details are included in this standard.

Table 9.2 Summary of HMM-Based DSV Studies

Research	Description
Dolfing [13]	Continuous HMM with Gaussian probability distribution
Kashi et al. [14]	Discrete HMM combined with a Mahalanobis distance of uncorrelated features
Rigoll et al. [15]	Discrete HMM using local features such as pressure, velocity, acceleration, and angle
Fierrez-Aguilar [16]	Continuous HMM using time functions of position, angle, velocity, and acceleration

At the international level the DSV standard is separated into two different parts, part 7 and part 11, of the ISO/IEC 19794 series. The ISO/IEC 19794-7 standard, published in 2007, specifies data interchange format of the x-coordinate, y-coordinate, pen angle, pen velocity, and pen acceleration in a time series representation. Part 11, still under development, specifies the data interchange of statistical features extracted from the raw signature.

9.7 Evaluations

The strength of DSV systems has been evaluated through various tests over the past couple of decades. In 1991 Sandia National Labs evaluated DSV systems as part of a larger test to compare the state of the art of different biometric technologies [17]. In 2004 the first International Signature Verification Competition (SVC) was conducted to compare multiple DSV systems against a common corpus of signatures [18]. The SVC was divided into two parts. A total of 15 international teams participated in first part, and 12 international teams participated in the second part. The data collected for the SVC contained the x-coordinate, the y-coordinate, the time series, the angle of the pen, altitude, and pressure for the signature attempts. Algorithms participating in part 1 used only the x-coordinate, the y-coordinate and the time series, whereas algorithms participating in part 2 could use all the characteristics. The lowest and the highest EERs for part 1 were 2.87% and 28.89%, respectively, and for part 2 they were 2.89% and 16.34%, respectively. The results showed that the additional information did not lead to better performance for the best algorithms, but it significantly helped the poorest performing algorithms.

Several large-scale signature databases have been collected as well with the intent of conducting technology evaluations. The Philips dataset contains signatures captured on a digitizer pad at a sampling rate of 200 Hz [13]. For each sampling point the x and y coordinates, the pen pressure, and the pen tilt angle were captured. Along with 51 genuine users, this dataset contains forgery attempts at three different effort levels. The first level includes forgeries from individuals who had an opportunity to observe genuine users sign in real time. The second level includes forgeries from individuals who were only given the resulting static signature. The third level includes forgeries from professionals in handwriting analysis.

The BIOMET database consists of data from five biometric modalities including signatures that were collected over three different sessions [19]. Approximately 130 subjects participated in the first session, 106 subjects participated in the second session, and 91 subjects participated in the third session. Data was collected using a Wacom Intuos2 A6 digitizer pad in the first session and a GripPen in the second and third sessions. At each sampling the x- and y-coordinates, pressure, azimuth, and altitude angles of the pen with respect to the digitizer pad were captured. Impostors attempted to sign for genuine individuals by forging samples of static signatures provided to them.

The MCYT database collected signature data as part of a larger biometric data collection in 2001 [20]. Data was collected using the Wacom Intuous A6 digitizer pad at a sampling of 100 Hz from a total of 330 users. At each sampling the x- and y-coordinates, pressure, azimuth, and altitude angles of the pen with respect to the digitizer pad were captured. In this dataset each genuine user was also asked to provide forgery attempts for five other users. Impostors were provided with the static signature for genuine users and asked to recreate the signature using their natural action. This dataset is available for academic and commercial research purposes.

The BioSecure Multimodal Database (BMDB) was collected from more than 600 subjects across 11 academic institutions in Europe [21]. Dynamic signature data was collected using the Wacom Intuos3 A6 digitizer tablet and the HP iPAQ hx2790 PDA at a sampling rate of 100 Hz. The users were spread across the academic institutions and data was collected in two separate sessions. Each individual provided 15 genuine signatures and 10 impostor attempts. The signature to be imitated was replayed on a screen and the forger could use that information for training. Two other datasets, BioSec and BioSecurID, were collected prior to the BMDB, and there is an overlap of modalities and users among the datasets. The BioSec database had 25 users in common with BMDB and BioSecuID had 31 subjects in common with BMDB. This data is intended for technology testing and interoperability testing.

Readers interested in a comprehensive review of DSV performance experiments should read [22].

9.8 Trends and Applications

Human signature verification has a long history in forensic analysis, although forgeries can go undetected by the untrained eye. DSV provides an additional level of confidence in the identity of the individual while maintaining the simplicity of the signing process. Due to its requirement of transactional security, the retail industry is a prime candidate for DSV. Every credit card transaction and other types of monetary transactions require a signature as form of authorization. Several large retail stores already store digital representations of signatures from credit card and debit card transactions. Incorporating DSV into these transactions can reduce fraud due to identity theft. Logical access to Web applications, network accounts, and computers can be authenticated using DSV. Biometric cryptosystems that derive cryptographic keys from DSV features have been developed as well.

9.9 Design and Deployment Considerations

Even with a significant body of research devoted to addressing DSV issues, there are several real-world considerations that need to be considered for a successful implementation.

9.9.1 Inherent Variability

The behavioral nature of DSV contributes significantly to variability of signatures. The positioning of the digitizer relative to the person signing has an impact on the final signature. For example, signing on a flat surface while sitting feels much different than signing on an angled surface while standing. This variability can affect the DSV characteristics. Other factors such as the psychological condition of the writer, fatigue, and injuries to the hand can also introduce variability in the signature action. Another factor that is relatively unexplored is the variability in signatures of an individual over an extended time period. Longitudinal studies examining this aspect are missing and need to be studied for its impact on performance of DSV systems.

9.9.2 Interoperability

The measurement of dynamic characteristics is dependent on the type of digitizer pad and stylus used for acquisition. Table 9.1 lists a number of digitizer pad technologies that introduce device-specific variations in the final signal captured from the user. Performance degradation is expected if the enrollment template is created on one type of a digitizer pad and verification is performed on another digitizer pad. Although the standards committee has addressed the issue of data interchange interoperability, sensor interoperability remains an unresolved issue.

9.9.3 User Demographics

The impact of user demographics such as gender, occupation, handedness, and age is still a relatively unexplored area. Research on the impact of age on DSV error rates has shown that performance is stable across age groups, but repeatability can be a challenge [23]. The research identified characteristics that can be incorporated into the system to improve performance for various age groups.

Individuals who are naturally left-handed tend to grip the pen differently than individuals who are right-handed. The signature surface is touch-sensitive and any additional interaction of the finger and the signature surface due to the grip type will introduce spurious features and negatively influence performance.

The impact of native writing language on DSV could prove to be a challenge as it is deployed in countries where English is not the native language. The current DSV systems are tuned for individuals used to writing in the English alphabet and although the matching algorithm should be language agnostic, other processes such as curve smoothing and quality analysis need to be studied in context of different languages.

9.9.4 Zero Effort Imposter Attempts

The testing of biometric systems requires imposter attempts for generating error rates, and there are several levels of imposter attempts based on the efforts of impostors. *Zero effort imposter attempts* are defined as an individual who presents his or her own biometric sample for verification against his or her own template, but the comparison is made against another individual's template. In DSV testing this means an impostor signs another person's name assuming that the signature will be compared against his or her own template. Thus, zero effort impostor attempts for DSV do not make sense. In traditional performance evaluation tests such as technology evaluations, imposter attempts are zero effort. This poses a challenge for performance evaluations, especially the calculation of FAR, of DSV systems. It also increases the difficulty of assessing the DSV system's ability to detect impostor attempts. The collection of DSV data for evaluations is an open research challenge, along with the determination of the forger and his or her skill level. Some research studies classify forgers into a skilled category if they have access to the original signature and have practiced signing the other signature [24]. In DSV the forger can be placed on a spectrum where the two extremes are zero effort and skilled, and in between imposters have varying amounts of information to conduct the forgery. Calculating a realistic FAR requires a certain level of effort on behalf of the impostor, and the level of effort has an impact on the ability to generate false accepts. Future research is required to clearly differentiate levels of forgery for DSV.

9.10 Summary

DSV technology is a natural extension of an action that is the most widely accepted form of authorization and thus attracts the least controversy of all biometric technologies. The signing process is intuitive, fast, and efficient, and DSV technology has shown that it can be used effectively for verification. With an increase in touch screens among the various devices used on a daily basis, DSV is well poised to increase its adoption in applications requiring transactional security. DSV has proven its accuracy potential through various academic and commercial research; future work needs to concentrate on interoperability and usability issues. Although DSV is unlikely to achieve wide-scale adoption like some other biometric technologies, specific segments such as the retail industry will likely be the early adopters of this technology due to its ability to increase transactional security without overhauling the entire customer interaction process.

References

[1] Kamins, D., and K. Zimmermann, "Signature Recognition Through Spectral Analysis," *IEEE International Conference on Acoustics, Speech and Signal Processing,* 1987, pp. 1790–1792.

[2] Herbst, N. M., and C. N. Liu, "Automatic Signature Verification Based on Accelerometry," *IBM Journal Research and Development,* Vol. 21, 1977, pp. 245–253.

[3] Sagar, N., *Unlocking Opportunities in the Healthcare and Government Sectors,* Frost & Sullivan, 2007.

[4] Osborn, A. S., *Questioned Documents,* Albany, NY: Boyd Printing Company, 1929.

[5] Plamondon, R., and G. Lorette, "Automatic Signature Verification and Writer Identification—The State of the Art," *Pattern Recognition,* Vol. 22, 1989, pp. 107–131.

[6] Johnson, R. R., and R. L. Dunham, "Personal Identification Method and Apparatus," U.S. Patent No. 3579186, 1971.

[7] Chainer, T. J., and T. K. Worthington, "Segmentation Algorithm for Signature Verification," U.S. Patent No. 4553258, 1985.

[8] Fierrez, J., and J. Ortega-Garcia, "On-Line Signature Verification," in *Handbook of Biometrics,* A. K. Jain, P. Flynn, and A. A. Ross, (eds.), New York: Springer, 2008, pp. 189–209.

[9] Nelson, W., and E. Kishon, "Use of Dynamic Features for Signature Verification," *IEEE International Conference on Systems, Man, and Cybernetics,* 1991, pp. 201–205

[10] Nelson, W., W. Turin, and T. Hastie, "Statistical Methods for On-Line Signature Verification," *International Journal of Pattern Recognition and Artificial Intelligence,* Vol. 8, 1994, pp. 749–770.

[11] Lee, L. L., T. Berger, and E. Aviczer, "Reliable Online Human Signature Verification Systems," *IEEE Transactions on Pattern Analysis and Machine Intelligence*, Vol. 18, June 1996, pp. 643–647.

[12] Garcia-Salicetti, S., et al., "Online Handwritten Signature Verification," in *Guide to Biometric Reference Systems and Performance Evaluation*, G. Dijana, et al., (eds.), New York: Springer, 2009, pp. 125–186.

[13] Dolfing, J. G. A., E. H. L. Aarts, and J. J. G. M. V. Oosterhout, "On-Line Signature Verification with Hidden Markov Models," *4th International Conference on Pattern Recognition*, Brisbane, Australia, 1998.

[14] Kashi, R., et al., "A Hidden Markov Model Approach to Online Handwritten Signature Verification," *International Journal on Document Analysis and Recognition*, Vol. 1, July 1998, pp. 102–109.

[15] Rigoll, G., and A. Kosmala, "A Systematic Comparison Between On-Line and Off-Line Methods for Signature Verification with Hidden Markov Models," *4th International Conference on Pattern Recognition*, Brisbane, Australia, 1998, pp. 1755–1757.

[16] Fierrez, J., et al., "HMM-Based On-Line Signature Verification: Feature Extraction and Signature Modeling," *Pattern Recognition Letters*, Vol. 28, December 2007, pp. 2325–2334.

[17] Holmes, J., L. Wright, and R. Maxwell, *A Performance Evaluation of Biometric Identification Devices*, Sandia National Laboratories, Albuquerque, NM, 1991.

[18] Yeung, D. -Y., et al., "SVC2004: First International Signature Verification Competition," *ICBA 2004*, 2004, pp. 16–22.

[19] Garcia-Salicetti, S., et al., "BIOMET: A Multimodal Person Authentication Database Including Face, Voice, Fingerprint, Hand and Signature Modalities," in *Audio-and Video-Based Biometric Person Authentication 2*, J. Kittler and M. Nixon, (eds.), Berlin, Germany: Springer Berlin/Heidelberg, 2003, p. 1056.

[20] Ortega-Garcia, J., et al., "MCYT Baseline Corpus: A Bimodal Biometric Database," *IEE Proc. Vision, Image and Signal Processing*, 2003, pp. 395–401.

[21] Ortega-García, J., et al., "The Multiscenario Multienvironment BioSecure Multimodal Database (BMDB)," *IEEE Transactions on Pattern Analysis and Machine Intelligence*, Vol. 32, 2010, pp. 1097–1111.

[22] Sayeed, S., et al., "Online Hand Signature Verification: A Review," *Journal of Applied Sciences*, Vol. 10, 2010, pp. 1632–1643.

[23] Guest, R., "Age Dependency in Handwritten Dynamic Signature Verification Systems," *Journal of Pattern Recognition*, Vol. 27, 2006.

[24] Kholmatov, A., and B. Yanikoglu, "Biometric Authentication Using Online Signatures," *International Symposium on Computer and Information Sciences 04*, LNCS, 2004, pp. 373–380.

CHAPTER 10

Keystroke Dynamics, Retina, DNA, and Gait Recognition

The previous chapters have discussed biological and behavioral biometric technologies that have an extensive amount of research and commercialization effort invested in them. There is no such thing as the perfect biometric technology. This chapter discusses three more biometric technologies that have garnered research and industry interest because they address the limitations of other technologies and take advantage of the existing infrastructure. Keystroke dynamics, DNA recognition, and gait recognition are discussed in this chapter due to their practical applicability in law enforcement and consumer applications. Several biometric technologies such as finger knuckle recognition, odor recognition, corneal topography recognition, and others are in their nascent stages are being investigated by research groups around the world.

10.1 Keystroke Dynamics

Computer systems are now pervasive in commercial, industrial, and individual activities. Businesses rely heavily on the effective operation of computer systems to ensure that their business run smoothly and successfully. The overwhelming majority of such computer systems use login-password combinations that rely on secrecy as a means of authentication. Password-based systems are susceptible to nontechnical attacks such as shoulder surfing, social engineering, and technical attacks such as brute force and dictionary attacks. Keystroke dynamics uses a person's typing rhythm for recognition. The uniqueness premise of keystroke dynamics stems from the observation that neurophysical factors that make written signatures unique also make typing patterns unique [1]. Keystroke dynamics has several inherent advantages: it does not require additional hardware, individuals are used to the authentication process, it is noninstrusive, and it does not require any deviation from the usual authentication process using passwords. Keystroke dynamics is currently used in the verification mode as an additional authentication layer on top of password secrecy.

10.1.1 History

Keystroke dynamics traces its roots to the telegraph and the Morse code. The U.S. military observed a phenomenon, later called the "Fist of Sender," whereby telegraphers could be identified based on the rhythm of the dots and dashes coming through in the telegraph. The RAND Corporation published a report in 1980 as part of a National Science Foundation (NSF)–funded project for authenticating individuals using keystroke dynamics [2]. Although this approach used an external device for capturing the timing of keystrokes, it proved that keystroke dynamics could be used to identify individuals in a 1:1 setting. The RAND experiment used seven secretaries and each secretary was asked to type three passages, each comprised of 300–400 words, at two different sessions held 4 months apart. The keystroke interval time was recorded for the experiment, and t-tests were carried out to check if the means of the keystroke interval times were the same at the two sessions. This experiment yielded encouraging results and brought attention to the potential of keystroke dynamics as a means of authentication. Another project funded by the National Bureau of Standards (NBS), now called the National Institute of Technology and Standards (NIST), in collaboration with SRI, concluded that keystroke dynamics based on username and password could provide a level of verification accuracy and a patent was awarded in 1989 [3].

Several keystroke dynamics applications have been commercialized in the last decade with varying degrees of success, with almost all applications focused on enhancing password-based authentication.

10.1.2 Feature Extraction and Matching

The mechanical action of typing provides two basic measurable features: keystroke press times and interkeystroke times, also referred to as keystroke latencies. The keystroke press time measures the time elapsed between pressing and releasing a particular key. The interkeystroke time measures the time elapsed between pressing and releasing successive keys. The measurement of interkeystroke times deserves a brief discussion. A keystroke action can be broken down into the press and release of a key. Thus, an interkeystroke time measurement can be described as any of the following four combinations (see Figure 10.1):

1. Key 1 press time to key 2 press time;
2. Key 1 press time to key 2 release time;
3. Key 1 release time to key 2 press time;
4. Key 1 release time to key 2 release time.

The selection of a specific combination for interkeystroke time is a design decision.

The secondary-level features of trigraph interkeystroke times have been used in previous research to provide additional discriminative power. Trigraph interkeystroke times measure the time elapsed between three consecutive keystrokes, and this concept can be extended to more than three consecutive keystrokes.

There are various approaches described in research literature for matching keystroke dynamics features that can be categorized into distance-based and

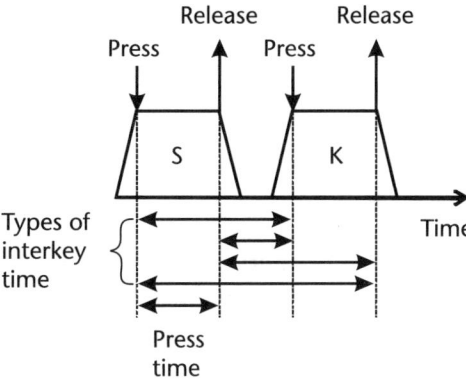

Figure 10.1 Keystroke features.

model-based methods. Distance-based methods use statistical parameters of the different features from the enrollment template and verification sample to determine if the difference between the two is within a specified tolerance. Various distance metrics such as the Euclidean distance and Mahalanobis distance measurements have been used in research. Model-based techniques use neural networks, support vector machines (SVM), and other methods to build a predictive model based on large amounts of training data from genuine and imposter users. Once the predictive model is trained, the verification sample is given as input and a decision is made by the predictive model. Section 10.1.5 discusses some prominent research studies conducted in this area.

10.1.3 Keystroke Dynamics Systems

Keystroke dynamics systems can be designed to recognize different typing inputs, although the basic measurable features of keystroke press time and interkeystroke time remain the same. The following sections discuss the different types of inputs.

10.1.3.1 Static Input

In this system an enrollment template is created for a specific word or phrase that is also used for verification purposes. This is the most common type of a keystroke dynamics system, as well as the most effective, as it integrates into the login-password process without disrupting it.

10.1.3.2 Continuous Input

In this system the typing rhythm is captured for an entire session and authentication is conducted throughout the session. In such a system authentication might have to be conducted on input not encountered in any previous sessions.

10.1.3.3 Application Specific Input

This system extends the concept of behavioral analysis to encompass typing patterns specific to the application. For example, an individual's typing pattern could change if he or she was writing a message in a social network application versus in a corporate e-mail application.

10.1.4 Standards

Currently, no standards exist for keystroke dynamics either from an application perspective or a data format perspective. A project proposal for creating a standard for keystroke dynamics data interchange was submitted to INCITS M1 in 2005, but it had not progressed to become a published standard as of the time of this writing.

10.1.5 Evaluations

Among the first published reports on keystroke dynamics were those by RAND Corporation in which seven subjects were used and a limited amount of typing samples were taken from each subject. Umphress and Williams conducted two different experiments that involved more subjects than the RAND experiment [4]. The first experiment included 17 programmers typing one passage of 1,400 words and another passage of 300 words. The 1,400-word passage served as a template and the 300-word passage was used as a verification attempt. The second experiment included 36 subjects typing a 530-word passage twice. The objective of these experiments was to study the mean of the keystroke interval timings for a larger population. The results showed that if five keystroke interval timings were used, the FRR was approximately 30% and the FAR was approximately 17%. The best results were obtained when using the keystroke interval timings for lowercase keystrokes with an FRR of 5.5% and an FAR of 5%.

Garcia introduced the term "electronic signature" in the patent for authenticating an individual [5]. His approach had the subjects type their own names, since there is less variance in the keystroke interval timings. First, participants typed their names a number of times, which created a template using the mean of the keystroke interval times and a covariance matrix of the keystroke interval times. When the individuals required verification, they would need to type their names and a verification feature vector would be created. If the verification vector was statistically similar to the template vector, then the attempt would be classified as an authentic attempt. Table 10.1 summarizes the results of some of the research in the past decade.

As part of its Comparative Biometric Testing Round 7, the International Biometric Group (IBG) tested a commercially available keystroke dynamics solution, from AuthenWare™ [6]; 212 subjects participated in this test and generated 4,851 genuine and 2,880 imposter attempts. An FMR of 3.264% and an FNMR of 3.195% were observed at a vendor-defined security level 3 on a scale of 1–5. Although the biological biometric technologies in this test outperformed keystroke dynamics, these results demonstrated the viability of keystroke dynamics as a cost-effective verification technology that strengthens password based mechanisms even if the password is obtained by an attacker. A benchmark dataset for keystroke dynamics is available for research purposes by Killourhy and Maxion [14].

Table 10.1 Summary of Keystroke Dynamics Research

Research	Classifier	Number of Users	FAR %	FRR %
Monrose and Rubin [1]	Bayes classifier	63	True verification rate = 94.14%	
Joyce and Gupta [7]	Distance: interkeystroke times	33	0.25	16.36
Obaidat and Sadoun [8]	Distance	15	0.7	1.9
	Neural network		0	0
Bergadano et al. [9]	Neural network	154	0.01	4
Yo and Cho [10]	Neural network	21	0	3.69
Boechat et al. [11]	Distance	24	0	4.44
Clarke and Furnell [12]	Neural network	32	EER = 5%	
Teh et al. [13]	Statistical	50	EER = 6.36%	

10.1.6 Trends and Applications

Password replacement and management are an area of immediate impact for keystroke dynamics, which is exactly on what most commercial systems are focused. The financial industry has been an early adopter of this technology where the adverse effect of password misuse can be extremely severe. Some of the earliest deployments of this technology were for intranet portal authorization for employees who have access to sensitive financial documents. Any information system that prefers using a password-based authentication mechanism while mitigating some of the risks associated with passwords make a good candidate for keystroke dynamics.

With the increasing trend of providing access to information using smartphones, keystroke dynamics could be seen on such devices in the near future. Smartphones work well with low processing and low power consumption applications, and keystroke dynamics offers a suitable solution. The feasibility of keystroke dynamics on smartphones still remains to be established, but there are ongoing efforts in this area.

10.1.7 Design and Deployment Considerations

The effectiveness of keystroke dynamics is influenced by factors that are based on users and the technology. Although there are factors that are specific to the application, the ones discussed next can be generalized to all types of applications.

10.1.7.1 Inherent Variability

Keystroke dynamics, unlike other physiological biometric technologies, is prone to change depending on the behavior of the person. Stress, fatigue, injury, posture, sitting or standing positions, and other factors can make a person's keystroke rhythm erratic. The steps of the authentication process should be carefully examined when considering keystroke dynamics.

10.1.7.2 Keyboard Effect

People learn their keystroke rhythms through repetitive mechanical action on a specific type of keyboard. A change in keyboard, either in size or layout, has a negative impact on matching capability. For networked systems where individuals can access logical resources using desktops, laptops, or mobile devices, interoperability becomes a prominent issue. Sensor interoperability is an issue that affects all biometric technologies—keystroke dynamics is no different.

10.1.7.3 Password Length

One method of countering the automated ways of breaking passwords is to increase the length of the password. Most password policies require users to use a password of a specific minimum length. A direct application of current password length policies to keystroke dynamics systems might provide the desired advantage. Commercial keystroke dynamics systems recommend passwords to be of a specific minimum number of characters, but research needs to be conducted to examine the trade-off between the length of the password and the decrease in the error rates of keystroke dynamics.

10.1.7.4 Timing API

Accurate and consistent time measurement is imperative for a keystroke dynamics system. Timing information can be recorded by using a variety of application programming interfaces (API) provided by programming environments such as JAVA and .NET or by using the operating systems functions. The *DateTime* object in .NET provides timing information accurate to within 10 milliseconds, whereas timing methods accessed by *kernel32.dll* provides a timing resolution up to 10 significant digits. The differences in timing resolution can have an impact on keystroke dynamics, although this has not yet been empirically tested.

10.2 Retina Recognition

The retina is located at the back of the eye and senses light and stimulates the optic nerve, as seen in Figure 10.2. The retina is comprised of blood vessels that form a unique pattern. It is the vascular pattern of the retina that is used for recognition purpose [15]. This pattern remains relatively stable throughout a person's lifetime and is well protected from external factors. The history of technology dates back to the 1930s when two ophthalmologists discovered that vein patterns of the retina are unique among individuals, which was further substantiated by studying the retina vein patterns of twins in the 1950s [16, 17]. Patents in retina recognition date

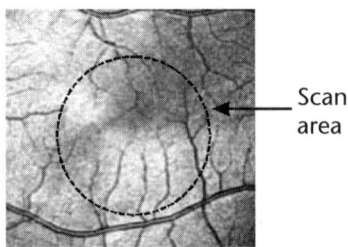

Figure 10.2 Retina image.

back to 1978 and in the 1980s a company called EyeDentify commercialized this technology for use in physical access control [18].

Retina acquisition requires scanning the inner wall of the eye and imaging a specific region in order to capture the retinal vascular pattern. The retina scanner illuminates the annular part of the retina through the pupil using infrared illumination or visible light. To ensure a high quality and consistent capture, alignment between the retina and the scanner is required. A chinrest is normally used in conjunction with a guide light on which the user focuses. This technology is very similar to that used by optometrists for eye examinations. The scanner progressively circles around a specific region of the retina and captures a circular retina image in which the vascular structure is visibly darker than the surrounding tissues. Several feature extraction and matching algorithms have been used for retina recognition, some of which are similar to the algorithms described in Chapter 8 [19].

Retina recognition has demonstrated a high degree of differentiation among individuals, but the acquisition mechanism is not user-friendly, which is its biggest limiting factor. Users consider the acquisition sensor to be intrusive and uncomfortable to use because they have to remove any eyewear, hold extremely still, and place their eyes very close to the sensor. Retina recognition was tested extensively in the 1980s, but has experienced extremely limited adoption and is unlikely to overcome that barrier unless the acquisition mechanism is made more user-friendly and cost-effective.

A unique application of this technology was demonstrated for the use in identifying cattle and sheep [20]. Animals also exhibit the same unique retinal features that can be used for recognition and a commercially available system for livestock retina recognition is available [21].

10.3 DNA Recognition

Deoxyribonucleic acid (DNA) recognition has been used extensively in forensic applications, and there is a significant body of knowledge to prove its ability to identify an individual. The human genome is coded on the 23 chromosome pairs present in every human cell and contains around 3 billion DNA base pairs. A small proportion of the DNA contains genes that contain instructions regarding various human biological functions. The rest is the noncoding portion of the DNA, which contains a highly repetitive sequence of nucleotides at different loci called *short tandem repeat* (STR). The analysis of these STRs in specific loci is used for recognition purposes. The common misconception is that a person's DNA structure is unique; rather, it is the DNA profile that is represented by the STRs at different loci that is relatively unique. Monozygotic twins exhibit exactly the same DNA profile, and similarity of STRs increases with the degree of relative consanguinity between individuals. This principle is used in DNA analysis for paternity tests. For biometric recognition, the number and loci of the STRs are defined by agencies conducting the DNA analysis, and the same STRs are analyzed for each individual. The Interpol Standard Set of Loci (ISSOL) specifies seven loci, while the FBI has selected 13 loci, which include the seven selected by Interpol [22].

DNA as a biometric has received a lot of interest in the last few years. DNA is well established in forensic sciences and has demonstrated high accuracy, but currently it faces several limitations for everyday use. It is monetarily expensive and time-consuming to conduct DNA analysis. Due to its nature of data collection and its use in law enforcement, DNA analysis is considered to be intrusive. The confidence in a DNA comparison is a function of number of loci analyzed and the degree of relative consanguinity [23]. Current techniques also require human examiners in order to make a final determination. DNA recognition can be highly accurate for unrelated individuals, but the probability decreases among related individuals.

All current operational DNA databases are used for law enforcement applications. In 1995 the United Kingdom launched its DNA database, and in 1998 the FBI created the Combined DNA Index System (CODIS) as a central repository for using DNA markers in law enforcement applications. The Interpol DNA Database, which has been online since 2006, is available to law enforcement agencies of all member states. The DoD Automated Biometric Identification System (ABIS) also has the capability of processing DNA information collected from subjects in the field. At the time of this writing, ISO/IEC SC 37 was developing a data interchange format, ISO/IEC 19791-14, for DNA data as part of the 19794 multipart standard.

There is a growing interest in reducing the current STR analysis time, which can be 8 hours or more, to about 1 hour using a mobile collection and analysis system. Currently, most of the interest is from law enforcement and intelligence agencies since they have access to the largest repositories of DNA profiles. Interested readers are encouraged to read [24].

10.4 Gait Recognition

Gait recognition technology uses the walking movement of an individual for recognition. The skeletal structure, the weight of person, and the person's acquired motor skills make a person's gait relatively unique. Gait recognition can be used on individuals at a distance and has garnered a lot of interest for surveillance applications. The earliest efforts in gait recognition can be traced back to early 1990s [25]. The HumanID Gait Challenge was initiated to create a gait dataset, set up various experiments for identifying technology gaps, and develop a baseline gait recognition technique. Gait is rhythmic action that repeats with each stride. The predominant approach for gait recognition is to capture a video sequence of the walking motion of an individual, create a silhouette representation, and extract the gait feat-ures. The rhythmic action influences the shape and movement dynamics of the individual, which is used for extracting features. Readers interested in technical details are encouraged to read [26]. Although identification using gait has shown promising results in highly controlled conditions, it is severely impacted by external factors such as change in weight, footwear, fatigue level, type of clothing, and contact surface. Results from the HumanID Gait challenge program exhibited a significant drop in identification accuracy when subjects changed their shoes between enrollment and the recognition sample. Gait recognition is still in its infancy and is likely to be used in combination with other biometric technologies to improve performance.

10.5 Summary

The biometrics domain is very dynamic, and several new technologies are being evaluated for its accuracy and convenience. Finger knuckle recognition and ear lobe recognition are a couple of examples of technologies that are being researched and developed for practical deployments. Out of all the technologies mentioned in this chapter, DNA recognition will continue receiving the most amount of attention in the near future. DNA recognition has a vast body of knowledge associated with it, but it will have to overcome usability challenges, public perception issues, and analysis limitations for it to become commercially viable as a biometric technology. The behavioral biometric technologies discussed in this chapter, keystroke dynamics and gait recognition, are excellent candidates for multibiometric recognition, which is discussed in Chapter 11. Keystroke dynamics and gait recognition might not provide the necessary discriminative power on their own, but their nonintrusive nature is advantageous to be used in combination with other recognition technologies.

References

[1] Monrose, F., and D. Rubin, "Keystroke Dynamics as a Biometric for Authentication," *Future Generation Computing Systems Journal: Security on the Web*, Vol. 16, 2000, pp. 351–359.

[2] Gaines, R., et al., *Authentication by Keystroke Timing: Some Preliminary Results*, 1980.

[3] Young, J. R., and R. W. Hammon, "Method and Apparatus for Verifying an Individual's Identity," 1989.

[4] Umphress, D., and G. Williams, "Identity Verification Through Keyboard Characteristics," *Int. J. Man-Machine Studies*, Vol. 23, 1985, pp. 263–273.

[5] Garcia, J., "Personal Identification Apparatus," 1986.

[6] IBG, *Comparative Biometric Testing Round 7 Public Report*, New York, 2009.

[7] Joyce, R., and G. Gupta, "Identity Authorization Based on Keystroke Latencies," *Communications of the ACM*, Vol. 33, 1990, pp. 168–176.

[8] Obaidat, M. S., and B. Sadoun, "Verification of Computer Users Using Keystroke Dynamics," *IEEE Transactions on Systems, Man, and Cybernetics*, Vol. 27, 1997, pp. 261–269.

[9] Bergadano, F., D. Gunetti, and C. Picardi, "User Authentication Through Keystroke Dynamics," *ACM Transactions on Information System Security*, Vol. 5, 2002, pp. 367–397.

[10] Yo, E., and S. Cho, "Keystroke Dynamics Identity Verification—Its Problems and Practical Solutions," *Computers & Security*, Vol. 23, 2004, pp. 428–440.

[11] Boechat, G. C., J. C. Ferreira, and E. C. B. Carvalho, "Using the Keystrokes Dynamic for Systems of Personal Security," *Proceedings of World Academy of Science, Engineering and Technology*, 2006.

[12] Clarke, N. L., and S. M. Furnell, "Authenticating Mobile Phone Users Using Keystroke Analysis," *International Journal of Information Security*, Vol. 6, 2007, pp. 1–14.

[13] Teh, P. S., et al., "Statistical Fusion Approach on Keystroke Dynamics," *Third International IEEE Conference on Signal-Image Technologies and Internet-Based System*, 2007, pp. 918–923.

[14] Killourhy, K., and R. Maxion, "Keystroke Dynamics—Benchmark Data Set," 2009.

[15] Patton, N., et al., "Retinal Image Analysis: Concepts, Applications and Potential," *Progress in Retinal and Eye Research*, Vol. 25, January 2006, pp. 99–127.

[16] Simon, C., and I. Goldstein, "A New Scientific Method of Identification," *New York State Journal of Medicine*, Vol. 35, 1935, pp. 901–906.

[17] Tower, P., "The Fundus Oculi in Monozygotic Twins: Report of Six Pairs of Identical Twins," *Archives of Ophthalmology*, Vol. 54, 1955, pp. 225–239.

[18] Hill, R. B., "Apparatus and Method for Identifying Individuals Through Their Retinal Vasculature Patterns," 1978.

[19] Borgen, H., P. Bours, and S. D. Wolthusen, *Visible-Spectrum Biometric Retina Recognition*, IEEE, 2008.

[20] Howell, B. M., et al., "Perceptions of Retinal Imaging Technology for Verifying the Identity of 4-H Ruminant Animals," *Journal of Extension*, Vol. 46, 2008.

[21] OptiBrand, "OptiBrand."

[22] *Interpol Handbook on DNA Data Exchange and Practice*, Lyon, 2001.

[23] Dessimoz, D., and C. Champod, "Linkages Between Biometrics and Forensic Science," in *Handbook of Biometrics*, A. K. Jain, P. Flynn, and A. A. Ross, (eds.), New York: Springer, 2008, pp. 425–459.

[24] Vallone, P. M., "DNA as a Biometric," *NSF Workshop on Fundamental Challenges for Trustworthy Biometrics*, Gaithersburg, MD, 2010, p. 14.

[25] Niyogi, S. A., and E. H. Adelson, "Analyzing Gait with Spatiotemporal Surfaces," *Proceedings of IEEE Workshop on Non-Rigid Motion*, 1994, pp. 24–29.

[26] Sarkar, S., and Z. Liu, "Gait Recognition," in *Handbook of Biometrics*, A. K. Jain, P. Flynn, and A. A. Ross, (eds.), New York: Springer, 2008, pp. 109–129.

CHAPTER 11

Multibiometric Systems

The majority of deployed biometric systems today use information from a single biometric technology for verification or identification. Large-scale identification systems have to address additional demands such as larger population coverage and demographic diversity, varied deployment environment, and more demanding performance requirements. Today's single modality biometric systems are finding it difficult to meet these demands, and a solution is to integrate additional sources of information to stregthen the decision process. A multibiometric system fuses information from multiple biometric traits, algorithms, sensors, and other components to make a recognition decision, which is discussed in Section 11.2. Research on multibiometric systems started in the 1970s when the future requirements of biometric systems indicated the need for low error rates and need to accommodate a larger part of the population. The last 5 years have seen an exponential growth in research and commercialization activities in this area, and this trend is likely to continue. This chapter will discuss design architectures, decision frameworks, opertional challenges, growth potential, and the future challenges of multibiometric systems.

11.1 The Need for Multibiometric Systems

Unimodal biometric systems face several challenges in today's implementations. The increasingly large enrollment population brings with it a range of issues such as missing biometric traits, the inability to provide good quality samples, and the refusal to use certain biometric traits due to religious and cultural concerns. For example, there is a certain subset of the population that is incapable of providing fingerprint images due to a genetic disorder called dermatopathia pigmentosa reticularis (DPR) [1]. Demographics and occupation have more of an impact on certain biometrics such as fingerprint recognition than others such as iris recognition. The capability of capturing another biometric trait can reduce the number of failure to enroll cases. Multibiometric systems are capable of capturing samples from multiple sensors. Environmental conditions have an impact on the ability of sensors and on the quality of captured data, and using multiple sensors increases the probability of acquiring good quality samples from at least one of the sensors. Spoofing of biometric systems is a growing concern, and a layered biometric system can improve security of the overall system. For a spoofing attack to be successful on a multibiometric system, all the biometric components would need to be successfully attacked. Multibiometric systems can be designed intelligently so that the matching performance of the system is better than a unimodal system. The multiple sources of information can be used to increase interclass variability and reduce intraclass variability. This is particularly useful for large-scale biometric systems, but this performance boost depends largely on the statistical independence of the biometric data, which is discussed later in this chapter. The decision process can be tuned at the individual level to give more weight to the better performing component of the multibiometric system. At a higher level, multibiometric systems provide additional information to resolve cases that are on the boundary of the decision policy.

11.2 Multibiometric System Design

There are several categories of multibiometric systems based on the type of input captured by the system or the number of different components used in the enrollment and recognition process. Figure 11.1 illustrates the different types of multibiometric systems.

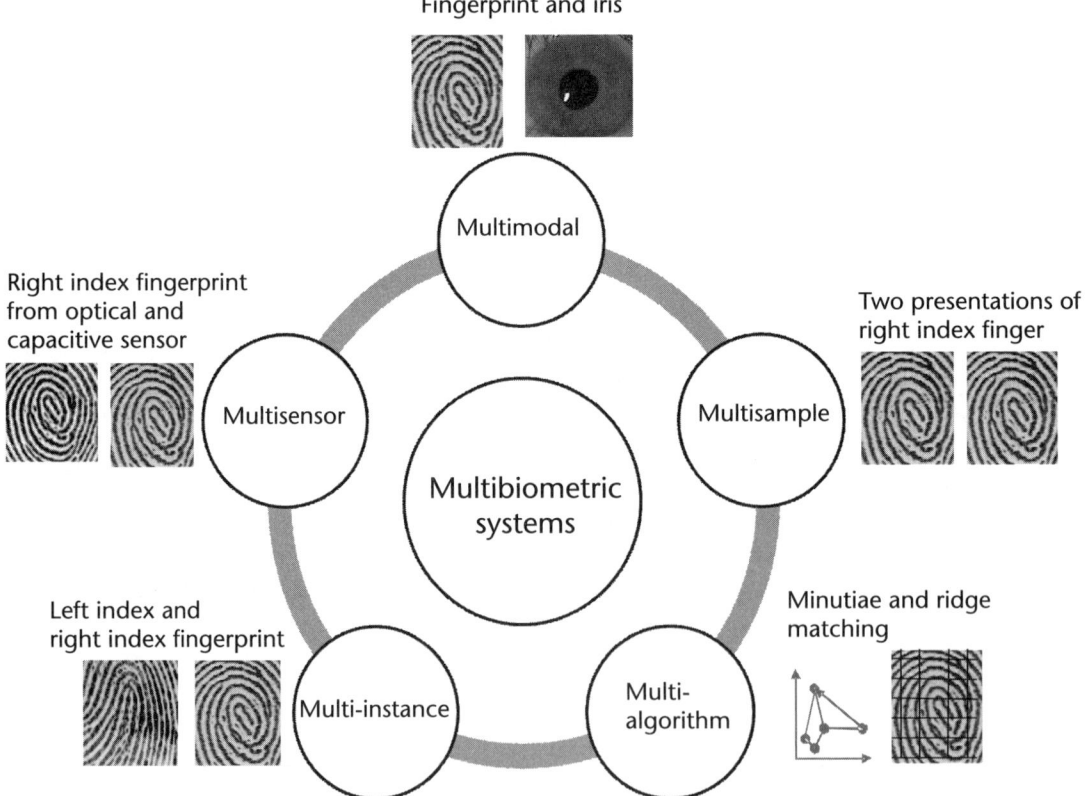

Figure 11.1 Types of multibiometric systems.

11.2.1 Multimodal

These systems fuse samples, features, or matching scores from multiple biometric technologies to make a decision and are the most common type of multibiometric systems. For example, a system that combines fingerprint recognition and face recognition is a type of multimodal biometric system. Such systems tend to be more expensive than other multibiometric systems, but are also the most common type of multibiometric systems.

11.2.2 Multi-Instance

These systems use multiple instances of the same physiological or behavioral characteristic as source of information. For example, iris images collected from the right and left irises of the same individual would qualify as a multi-instance system. The US-VISIT program, which uses a 10-print fingerprint sensor to collect fingerprints from multiple fingers, is an example of an operational multi-instance system.

11.2.3 Multisensor

These systems use multiple sensors to collect the same physiological or behavioral characteristic. Such systems are capable of capturing different levels of information from the source without any additional interaction between the subject and the sensor. For example, a face system consisting of separate cameras acquires face images using visible light and infrared illumination, which results in the acquisition of complementary information of the face light at different wavelengths. This information is then combined to gather more details of the face than a single type of camera.

11.2.4 Multialgorithmic

These systems process the same physiological or behavioral traits using multiple feature extractors or feature matcher. These systems use a single sensor to capture the biometric trait and combine information from two or more extractors or matchers to form a decision. A single sensor and a single presentation reduce the number of human interaction issues in such systems.

11.2.5 Multisample

These systems capture multiple samples of the same biometric trait. The interaction between the subject and sensor invariably distorts the acquired sample. Combining multiple samples of the same trait helps reduce the inherent variability in the resulting feature vector. Some commercial systems capture multiple samples to create an enrollment template, which is also sometimes referred to as a generalized template. Such templates have demonstrated an improved recognition performance compared to single sample templates. Several vendors recommend a multisample enrollment template as a best practice.

11.3 Data Acquisition

The presentation of biometric traits to the sensor can be performed in two different ways: *sequential* and *simultaneous* (see Figure 11.2) [2]. In sequential acquisition the biometric traits are captured one at a time, and the successive data capture cannot start unless the previous one has completed. Sequential acquisition is more time-consuming and requires multiple interaction events that can introduce variability and inconsistency in the sample, as well as increase inconvenience to the user. Simultaneous acquisition requires a single interaction event and is quicker to perform. Such an acquisition could require specialized hardware capable of capturing all the details in a single event. For example, a sequential capture of the face and iris would require the user to first provide his or her face image to a camera, disengage, and then provide his or her iris image to an iris camera. A simultaneous capture of face and iris images would require that the hardware is integrated or that the separate devices are capable of capturing the biometric samples at the same time.

11.4 Levels of Fusion

A biometric system is comprised of five subsystems: data capture, signal processing, data storage, matching, and decision making. Multiple levels of fusion can be categorized based on the subsystem responsible for fusing the information. These are discussed next.

11.4.1 Sensor Level Fusion

The raw biometric sample is the richest source of information since it has been minimally processed. Sensor level fusion combines the raw digital signals from two or more samples and sends a single, combined sample to the feature extraction module, as shown in Figure 11.3. These samples could be multiple acquisitions of same biometric trait from a single sensor or multiple acquisitions of different traits. Sensor level fusion is simpler for multiple acquisitions of the same biometric trait since the signal representation is similar in all the samples. The consolidation of multiple samples from different biometric traits requires an in-depth knowledge of

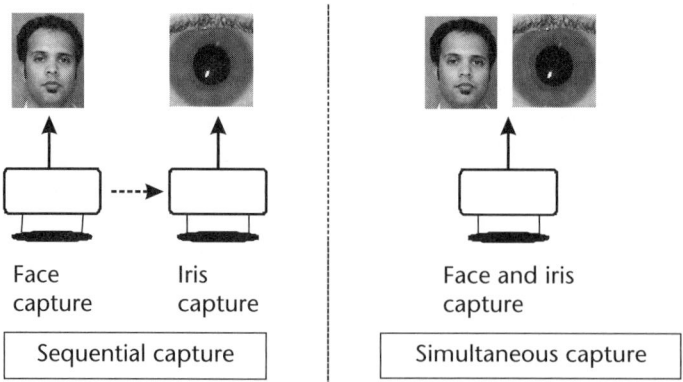

Figure 11.2 Sequential and simultaneous data capture.

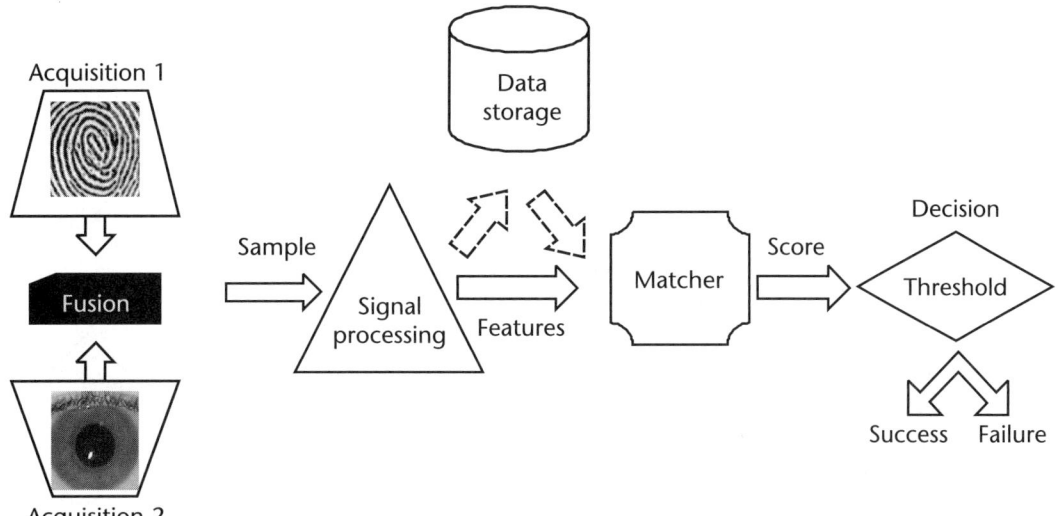

Figure 11.3 Sensor level fusion.

the signal space and is not possible for samples that have inherently different representations, for example, fingerprint images and voice samples.

11.4.2 Feature Level Fusion

The signal processing subsystem extracts features of interest from the signal for storage and matching purposes (see Figure 11.4). This fusion level combines multiple feature sets and creates a single template. For multibiometric systems where the samples are not independent, the dependencies in the feature sets can be used to improve consistency and detail. The simplest form of feature level fusion is to concatenate two feature vectors, but in practice this is challenging. The various biometric feature sets may be incompatible or have an extremely high level of dependency between the feature sets that can suppress important details. Also, most commercial systems do not provide access to the raw feature sets. Although research has explored combining feature sets of different modalities, most practical applications will attempt to use this level of fusion for multi-instance, multisensor, or multisample systems.

11.4.3 Score Level Fusion

The match score is calculated based on the degree of similarity between two biometric samples (see Figure 11.5). In score level fusion multiple scores from multiple sources are integrated to generate a single match score. Score level fusion is the most researched topic in the multibiometric domain as match scores are easily available from biometric systems and a detailed understanding of the matcher's behavior can be observed through large-scale testing. Theoretically, it is possible to create formal methods of validating performance improvement of a particular score level fusion based on underlying matching score density functions. The effectiveness of score level fusion techniques increases if accurate information about the score range and

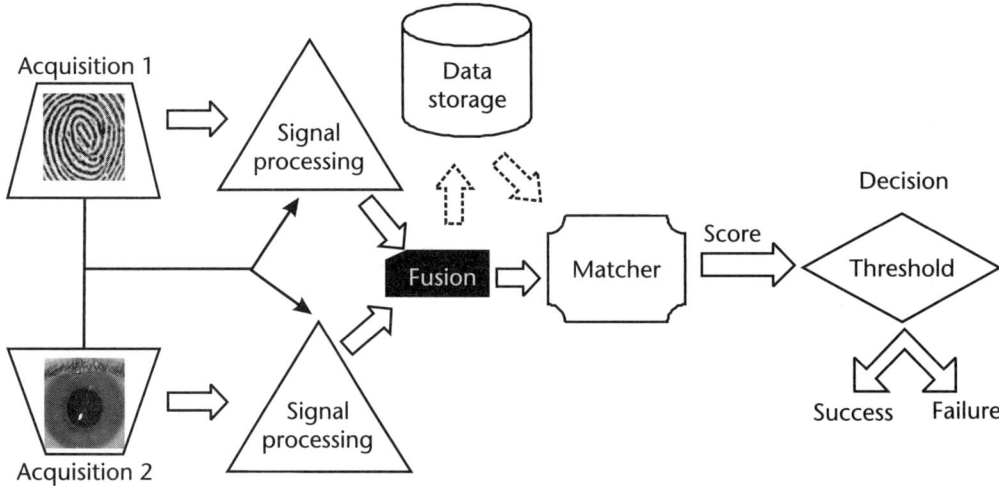

Figure 11.4 Feature level fusion.

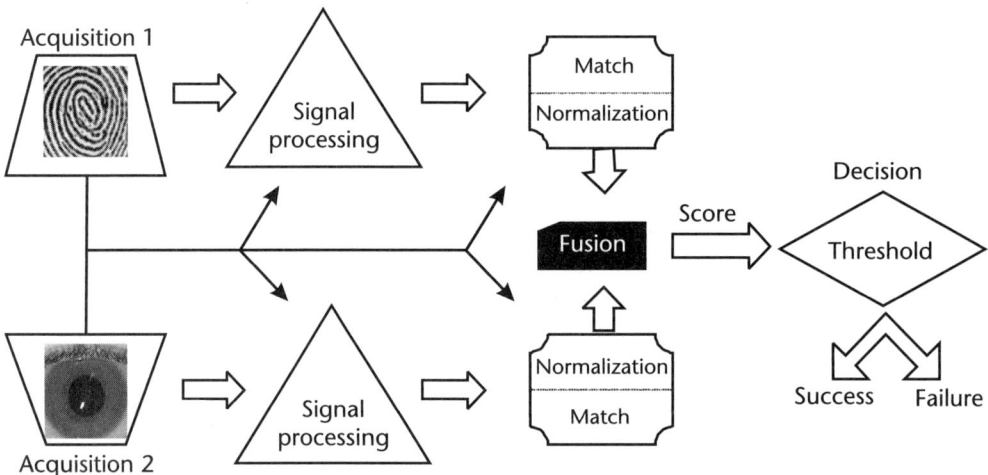

Figure 11.5 Score level fusion.

performance parameters of the underlying matcher is available. Score level fusion techniques are generally categorized into two groups: classification approach and combination approach. These are discussed in Sections 11.4.3.2 and 11.4.3.3 and a summary of the techniques is provided in Table 11.2.

Certain score level fusion techniques require transforming the match scores so that they can be interpreted on a common scale. For example, if matcher X produces a score on a scale of [1, 100] and matcher Y produces a score on a scale of [1, 2,500], then how does a fusion method combine the scores without losing the confidence exhibited by each matcher's score? In such cases the scores are transformed to map them to a common domain. Certain matchers that generate scores of different data types such as interval or ratio can also require score normalization. Score normalization is an extensive area of research onto itself, which is discussed in Section 11.4.3.1.

11.4.3.1 Score Normalization

The process of score normalization requires changing the parameters of matchers and data types to map matching scores to a common domain. An example of normalization is shown in Figure 11.6. Traditionally, score normalization methods have been evaluated on the basis of robustness and efficiency. Robustness represents the impact of outliers on normalization, and efficiency represents the proximity of the calculated normalized estimate to the optimal estimate when the underlying distribution is known. The genuine and imposter score distributions, as well as empirical data, can be used to determine the robustness and efficiency of various methods. Readers interested in learning the technical details of score normalization are encouraged to read [3]. Some of the more commonly used score normalization methods and their characteristics are discussed in Table 11.1 [4].

11.4.3.2 Classification Approach

Classification methods formulate the problem as dividing the decision space into two classes: genuine and imposters. Classifiers require large amounts of training data to determine its optimal decision boundary, which ultimately determines its effectiveness. The reliability and precision of the decision boundary is dependent on the amount and quality of input data available for training the classifier, which is a limitation of this approach. However, matching scores need not be homogenous and thus normalization is not required. An input feature vector is created by combining the matching scores from the different sources. During the training phase, or enrollment, input feature vectors representative of the genuine users and imposters are used. During recognition the input feature is classified and a decision is made. Neural networks, nearest neighbor algorithms, and tree-based classifiers are some of the techniques that have been researched with varying levels of success.

11.4.3.3 Combination Approach

This is the most common and computationally effective method for match score fusion. This method combines match scores from multiple sources and calculates a

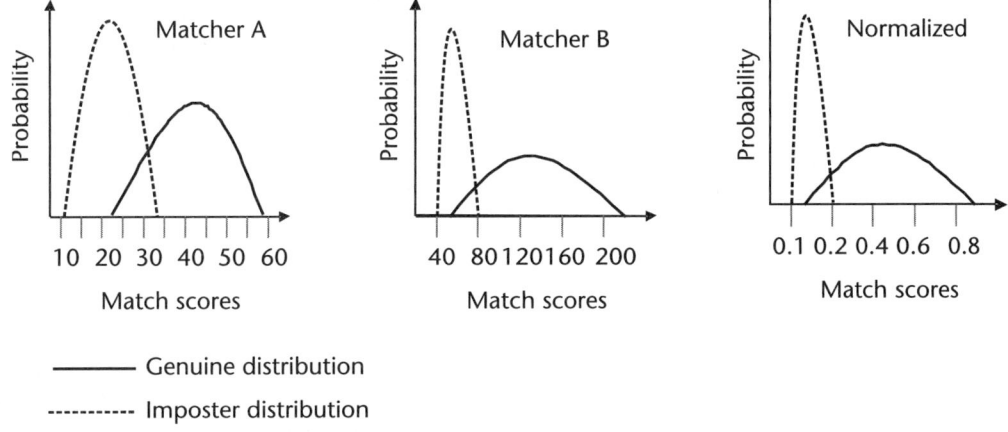

Figure 11.6 Score normalization example.

Table 11.1 Score Normalization Methods

Name	Characteristics
Min-Max	Best suited for known score ranges; sensitive to outlier data
Decimal Scaling	Assumes scores on a logarithmic scale
Z-Score	Assumes normal distribution and symmetry around the mean
Median absolute deviation (MAD)	Insensitive to outliers and extreme values; performs best on normal distribution
Tanh estimators	Insensitive to outliers in case of few points on the extreme tails of distribution; for tail-heavy distributions, parameter selection influences efficiency
Adaptive methods: two quadratics, logistic, quadratic-line-quadratic	Assumes nonlinearity of data; uses iterative normalization process to increase separation between genuine and imposter distributions

Source: [4].

Table 11.2 Score Fusion Techniques

Method	Description
Combination Approaches	
Simple sum	No probability distribution or matcher parameters required
Minimum score	No probability distribution or matcher parameters required
Maximum score	No probability distribution or matcher parameters required
Matcher weighting	Genuine and imposter probability distributions
User weighting	Individual performance parameters
Sum of probabilities	Genuine and imposter probability distributions
Product of probabilities	Genuine and imposter probability distributions
Likelihood ratio	Genuine and imposter probability distributions
Classification Approaches	
K-nearest neighbor	Genuine and imposter match scores
Decision trees	Genuine and imposter match scores
SVM	Genuine and imposter match scores
Neural network	Genuine and imposter match scores
Discriminant analysis	Genuine and imposter match scores

Source: [3].

single match score. This approach requires normalization of match scores to aid in the fusion process. A theoretical framework for a combination approach was proposed and evaluated by Kittler et al. [5]. This framework requires the conversion of matching scores from the various classifiers into posteriori probabilities of genuine and imposter distributions and then applies the product rule, sum rule, min rule, max rule, and other rules. The effectiveness of these rules is based on assumptions regarding the statistical independence of data and proximity of posteriori probabilities to priori probabilities. Other combination approaches use simple sum, min score, max score, and weighting rules to consolidate matcher scores. The efficacy of these rules was evaluated in [6].

Methods based on Bayesian classifiers are popular where posterior joint probability densities of multiple modalities are not available for genuine and imposter distributions. The Bayesian classifier can be applied by calculating the posterior probability density for each modality and combining the individual decisions using the product rule to make a final decision.

11.4.4 Rank Level Fusion

Biometric systems that operate in the identification mode are capable of returning candidate lists that are in descending order of similarity with the input sample. Rank level fusion is used to combine ranks provided by multiple matchers or classifiers to make an identification decision. In this method the amount of information available to the fusion module is less than score level fusion, but the comparability of the candidate lists does not necessitate normalization, which reduces its computational complexity. Various rank level fusion techniques have been developed [7, 8]:

1. The *highest rank fusion method* assigns the highest rank from among multiple matchers to the identity. The effectiveness of this method increases where a large number of matchers are available as the probability of a tie occurring increases as the number of matchers decreases.
2. The *Borda count fusion method* uses the aggregate ranks of all matchers to produce a final rank. The effectiveness of this method is determined by statistical independence of the matchers and relatively similar performance metrics of all matchers.
3. The *weighted Borda count fusion method* uses logistic regression to generate weights for individual ranks produced by all the matchers, which is then used in the Borda count fusion method.
4. The *Bucklin majority vote fusion method* uses a majority vote on the individual ranks to generate a final consolidated rank.

Rank level fusion has received relatively less attention than score level fusion and is expected to generate more interest with proliferation of identification systems.

11.4.5 Decision Level Fusion

This level integrates decisions generated by multiple biometric systems to make a final determination. From a practical perspective, this is the simplest type of fusion as all biometric systems provide this information. From an information perspective this provides the least amount of detail as the output of a decision subsystem is always binary: match or nonmatch. Several decision fusion techniques have been studied by researchers. The "AND" and "OR" logic rules can be applied effectively to decision fusion. Voting schemes based on majority and weighted majority rules have been established as well. The Bayes classifier, discussed earlier in score level fusion, can also be used for decision level fusion. The classifier approach requires a large amount of training data to establish the underlying accuracy of the individual decision modules.

As the fusion level progresses from the sensor level to the decision level, the amount of information available deceases along with the computational complexity of the fusion methods. Although the impact of this relation has not been quantified in terms of performance rates, the current state of technology knowledge and development has made score level fusion the most researched area.

11.4.6 Quality-Based Fusion

Incorporating quality scores into the fusion process is a relatively new and unexplored area in multibiometric systems. By contrast, quality assessment in unimodal biometric systems is understood to a higher degree, as exhibited by predictive results produced by NFIQ for fingerprints. Within multibiometric systems, quality assessment is a more complicated process. The primary challenge is to assess the utility of unimodal biometric technologies on the various fusion frameworks of multibiometric systems since this requires understanding the combination effect on the fused sample, feature vector, score, or decision. Various techniques are being researched to generate quality scores with a high degree of performance predictability for multibiometric systems. Most of this research is concentrated on calculating utility for score fusion frameworks. Within the context of multibiometric systems, quality scores are used to assign weights to fusion modules or customize the invocation order of the different components in a multibiometric system. Quality scores have been used to modify the training process for classifiers such as SVM and Bayesian classifiers [9, 10]. Quality scores have also been used to estimate the joint density of match scores and to assign weights to individual matchers using the likelihood ratio test. These matcher weights are then used during score level fusion. Research conducted on iris images showed a genuine accept rate (GAR) at a fixed FAR of 0.01% of 75.2%, 89.5%, and 94.8% for single iris images, product fusion rule, and quality-based product fusion rule, respectively [11]. A quality score based on the utility of multibiometric samples is an effective means of improving performance and is likely to receive increased attention.

11.5 Standards

In 2007 ISO/IEC SC 37 published a technical report titled *Multimodal and Other Multibiometric Fusion,* which describes various aspects of multibiometrics and identifies standardization challenges [2]. ISO/IEC 29159-1:2010, titled "Biometric Calibration, Augmentation and Fusion Data—Part 1: Fusion Information Format," addresses the interoperability and standardization of score fusion data in multibiometrics [12]. Certain score fusion techniques require score normalization to transform the scores to a common domain. Score normalization is aided by statistical data related to genuine and imposter distributions, and this standard defines a fusion information format (FIF) to support popular transformation techniques. By standardizing the data format required for normalization, the individual matchers and fusion algorithms can remain proprietary. For example, a fingerprint and face recognition system will provide the matching score for each modality along with genuine and imposter distributions for each matcher in the FIF, which will then be provided to the fusion module for further processing.

The area of performance testing and reporting is likely to be most active and challenging areas of standardization for multibiometric systems. Current testing and reporting standards are targeted at unimodal biometric systems that do not specify requirements for multibiometric systems. At the time of this writing, a new work item was proposed in SC 37 for evaluating and reporting the performance of multibiometric systems. The proposed standard will include a description of the performance measures of multimodal biometric devices, a specification of the biometric data collection and performance calculation, and a specification of the reporting performance.

The standardization of technical interfaces and application profiles for multibiometric systems is also expected in the near future. The BioAPI framework is currently capable of processing a single type of modality, even though it is modality-agnostic. Different frameworks may be required for various types of fusion techniques. Similarly, application profiles that use multibiometric systems will require the specification of fusion algorithms for the application, confidence levels for the fusion process, and types of biometric modalities allowed.

11.6 Evaluations

Jain et al. [6] conducted an evaluation using three fingerprint recognition systems and a face recognition system, which were all commercially available. They used fingerprint data from 972 individuals collected on a 500-dpi optical sensor and associated it with 972 images from the FERET database, thus creating a virtual multimodal dataset for 972 unique individuals. The EER was evaluated for all combinations of four normalization methods and five match score fusion methods. The overall results, listed in Table 11.3, showed that multimodal combinations performed better than each unimodal system, but the performance improvement was not significantly large. The simple sum rule, which is an extremely computationally efficient and easy-to-implement method, performs relatively similar to some of the other more complex score fusion techniques.

A study on multimodal 2-D and 3-D face recognition was released in 2005 [13]. Approximately 198 subjects participated in the study and provided 3-D face images under a single capture condition and 2-D faces images under four different capture conditions. A sum rule based on linear transformation was used for fusing scores of the multisample 2-D and 3-D face images. The results showed that EER for 2-D+3-D multimodal fusion was 1.9% compared to 4.3% for unimodal 2-D face recognition and 4.8% for unimodal 3-D face recognition.

A report titled *Studies of Biometric Fusion* was released by NIST in 2006 and contained results on a variety of score level fusion studies using fingerprints and

Table 11.3 EER (%) for Combination of Normalization and Score Fusion

Normalization Methods	Fusion Methods				
	Simple Sum	Min. Score	Max. Score	Matcher Weighting	User Weighting
Min-Max	0.99	5.43	0.86	1.16	0.63
Z-Score	1.71	5.28	1.79	1.72	1.86
TanH	1.73	4.65	1.82	1.50	1.62
Quadratic-Line-Quadratic	0.94	5.43	0.63	1.16	0.63

Source: [6].

face images of 187,000 individuals [14]. The study had two distinct objectives: assess the performance of eight different score fusion techniques, and assess the score level fusion on multimodal, multi-instance, multisample, and multimatcher architectures. Three best-performing fingerprint and face matchers from previous NIST evaluations were used in this study. Out of the eight techniques listed in Table 11.4, the product of likelihood ratios and logistic regression were found to be the most effective.

In the second part of study, the product of likelihood ratio technique was used to assess the improvement of FRR in six different multibiometric scenarios. The results are listed in Table 11.5.

Out of all the methods, multimatcher score level fusion was found to be the least effective, whereas other methods that required multiple samples or instances were more effective. This indicates the positive impact of data independence, in the case of the finger and the face, and data variability, in the case of the multisample. The study also highlighted the need to understand the underlying probability distributions and score correlations for implementing effective score fusion methods.

In an experiment conducted by Snelick et al. [15] on score fusion techniques of more than 1,000 subjects using commercially available face and fingerprint recognition systems, four score normalization techniques (Min-Max, Z-score, TanH, MAD) and five score fusion techniques (simple sum, max score, min score, sum of probabilities, product of probabilities) were evaluated. The GAR of fingerprint

Table 11.4 List of Score Fusion Techniques

Method	Data Required
Product of likelihood ratio	Univariate genuine and imposter distributions
Logistic regression	Univariate genuine and imposter distributions
Product of FARs	Univariate imposter distribution
Best linear	Joint genuine and imposter distributions
Sum of Z-norm scores	Univariate imposter distribution
Max. of FAR	Univariate imposter distribution
Min. of FAR	Univariate imposter distribution
Sum of raw scores	None

Source: [14].

Table 11.5 Improvement in FRR at FAR of 0.01%

Method	%	Matcher 1	Matcher 2	Matcher 3
10 fingers and 3 face matchers	Max	80	84	79
	Average	74	77	74
Two fingers	Max	90	90	84
	Average	82	78	71
Multisample (fingerprint) fusion	Max	70	72	57
	Average	N/A	N/A	N/A
		Matcher 1 + 2	Matcher 1 + 3	Matcher 2 + 3
Multiple matchers (fingerprint)	Max	33	32	32
	Average	25	16	20

Source: [14].

recognition at the FAR of 0.1% was 83% and the GAR of face recognition at the FAR at 0.1% was 75.3%. The combination of min-max normalization and simple sum score fusion demonstrated the most improved performance of the GAR of 98.7% at the FAR of 0.1%. The combination of other normalization and score fusion methods also performed better than both the unimodal systems.

Data collection protocols for multibiometric systems are an area that has not received much attention. One of the larger efforts in this area is the BioSecure Multimodal Database, which was collected by 11 institutions in the European Union and is one of the largest multimodal and multiscenario databases collected for research purposes [16]. More than 600 participants provided fingerprint, signature, hand, iris, face, and voice data in different scenarios. The first scenario collected face and audio data over the Internet. The second scenario collected fingerprint, signature, iris, hand, face, and audio data using a desktop PC. The third scenario collected fingerprint, signature, face, and audio data using a mobile device. This database is available to researchers through the BioSecure program.

There are several evaluation databases available for research purposes that are discussed in depth in Chapter 13.

11.7 Trends and Applications

A multibiometric system, because of the nature of the problems that it is trying to solve, is better suited to large-scale identity management systems such as national ID programs and border control applications. The Unique Identification Authority of India (UIDAI) has initiated a project to provide all Indian residents, on a voluntary basis, currently numbering around 1.2 billion, with a unique 12-digit number. This unique number will be associated with the user's 10 fingerprint images, two iris images, and a face image. This is an example of a multimodal and multi-instance type of system. The exact details of how all this information will be used in the decision process is not known, but it is likely that it will be used in the de-duplication effort during the enrollment process. During the verification process, the user will be given a choice of biometric technology.

The Biometric Automated Toolset (BAT) used by the U.S. military in Iraq and Afghanistan is a successful real-world deployment of multibiometrics. The BAT system includes a laptop, a fingerprint scanner, an iris scanner, a camera, and an ID card printer. The BAT system is used to create records of residents, wanted individuals, and detainees and it shared across multiple military posts across Iraq. This allows a biometric identification check of individuals when they move from one region to another and a determination of their civilian status.

The Next Generation Identification (NGI) program being developed by the FBI will replace the current IAFIS program. One of the key goals of this program is to provide the capability of integrating multimodal biometric technologies into the new system. Although fingerprint recognition will still serve as the basis of all matching operations, it is likely that iris recognition will be used increasingly in NGI.

Consumer-facing applications in segments such as retail industry, healthcare industry, and financial institutions are unlikely to adopt multibiometric systems in its true form because their operational environments still do not demand such systems. Such segments would still prefer to use unimodal biometric systems, and a practical solution to the universality issue would be to have different types of operational biometric systems and use specific systems based on user preference.

11.8 Design and Deployment Considerations

Some of the challenges facing multibiometric systems have already been discussed in earlier sections. Multibiometric systems are gaining momentum due to several reasons, key among them the performance improvement and accommodation of a larger number of subjects. The type of multibiometric system will depend on the eventual aim of the system: increase accuracy or increase coverage. Other influencing factors are discussed next.

11.8.1 Cost

The choice and number of biometric traits is largely driven by the level of security required, the type of application, and the budget. The acquisition cost of a multibiometric system increases with the number of technologies used. The maintenance and administrative costs are also linked to the number of biometric traits and the type of multibiometric system used.

11.8.2 Correlation

The technical performance of the system depends on the statistical independence of the parameters used in the fusion technique. For example, a multialgorithm system that uses the same fingerprint is likely to have a higher correlation of results since the same information is used by all the algorithms. The Iris Challenge Evaluation (ICE) 2005 conducted by NIST indicated a correlation between average match and nonmatch scores of the left and right iris images of the test subjects [17]. This result implies that score fusion using two iris images of the same individual, even though completely different, may not provide the same results as score fusion from independent iris images. A multi-instance system that uses four fingers is likely to have a lower correlation of results as the more information is available to the system. Theoretically uncorrelated data is preferable since it provides additional information to the fusion process, but this has to be supported by empirical data as well as it may not be true for all biometric traits. Correlation needs to be considered in all levels of fusion and fusion architectures. The uncertainty regarding accuracy improvements is an area that is actively being researched and needs to be addressed to increase consumer confidence.

11.8.3 Human Factors

For certain types of multibiometric systems, such as multimodal, multi-instance, multisample, and multisensor, human interaction challenges can be more complex than unimodal systems. The point of fusion in the multibiometric system and the nature of presentation (*sequential* or *simultaneous*) determine which human factor issues need to be addressed. The latest iris cameras are capable of acquiring the left and right irises in the same interaction, whereas the older generation cameras required a user to provide one iris image, disengage, and then provide the other iris image. The human factor challenges are quite different in these two systems even though they are both multi-instance systems. Multibiometric systems using simultaneous data capture requires that the sensors are tightly integrated, which can directly impact the final cost of the system. The enrollment and recognition processes need to be tailored to the type of data acquisition, thereby affecting the role of the system operator.

11.8.4 Fusion Architecture

Which type of a multibiometric system and fusion strategy is the best? This is a question that several researchers have attempted to answer, but the answer depends on factors such as operational environment, availability of matcher information, demographics of target population, and operational requirements. Use-case specific information is not available yet and most real-world deployments have to rely on research projects results generated in a laboratory environment.

11.8.5 Quality of Samples

The impact of sample quality on performance of unimodal systems is a heavily researched area. How to effectively use this information in multibiometric systems is a relatively nascent effort. Feature level fusion techniques require a completely new formulation of quality metrics. Score level fusion techniques require an understanding of the level of correlation between the feature sets and quality scores to effectively use quality metrics. The question of how quality scores should impact the fusion process remains unanswered and is a pressing issue that deployed systems need to address.

11.8.6 Zoo Analysis

Chapter 2 discussed user-specific performance in terms of zoo analysis. Applying the same concept to multibiometric systems raises an interesting issue: Does a user belong to the same zoo category as they transition from one biometric technology to another? For example, if a user is a sheep for a fingerprint recognition system, is there a way to determine the user's zoo category for a face recognition system? Empirical data analysis can provide insight on this question, and a pilot study using a small sample of users should be conducted to understand how the target population will behave in a multibiometric system.

11.8.7 Privacy

Privacy protection is the biggest nontechnical challenge facing biometrics. Multibiometric systems capture more information than unimodal biometric systems, and in most cases they capture information from different traits. A multimodal system that captures fingerprints, face images, and iris images now becomes an even bigger target for attackers and raises greater concerns among users about misuse of their biometric data. Appropriate security measures and privacy policies are a must for all biometric systems, and they carry even more importance in multibiometric systems.

11.9 Summary

Multibiometric systems are the new frontier and this is evident from the number of ongoing efforts in the research and commercial domains. They are an answer to the deficiencies of unimodal biometric systems, namely, the improvement of performance and the reduction of failure to enroll rates. Theoretically, they represent a huge potential, but research and commercial systems have yet to reflect this promise. The lack of standards also indicates that more work is required before it reaches an acceptable level of maturity. Current operational multibiometric systems are designed to capture multiple traits and store the raw data separately and use it in a layered decision process, and this practice is unlikely to change in the near future. Multibiometric systems can use existing technology and improve the performance of large-scale databases, and these advantages will drive the development of multibiometic systems.

References

[1] Handwerk, B., "Born Without Fingerprints: Scientists Solve Mystery of Rare Disorder," *National Geographic News*, 2006.

[2] ISO/IEC, *ISO/IEC TR 24722:2007 Information Technology—Biometrics—Multimodal and Other Multibiometric Fusion*, Geneva, Switzerland, 2007.

[3] Ross, A. A., K. Nandakumar, and A. K. Jain, "Score Level Fusion," in *Handbook of Multibiometrics*, New York: Springer, 2006, pp. 91–141.

[4] Jain, A. K., K. Nandakumar, and A. A. Ross, "Score Normalization in Multimodal Biometric Systems," *Journal of Pattern Recognition*, Vol. 38, 2005, pp. 2270–2285.

[5] Kittler, J., et al., "On Combining Classifiers," *IEEE Transactions on Pattern Analysis and Machine Intelligence*, Vol. 20, 1998, pp. 226–239.

[6] Snelick, R., et al., "Large-Scale Evaluation of Multimodal Biometric Authentication Using State-of-the-Art Systems," *IEEE Transactions on Pattern Analysis and Machine Intelligence*, Vol. 27, March 2005, pp. 450–455.

[7] Ho, T. K., J. J. Hull, and S. N. Srihari, "Decision Combination in Multiple Classifier Systems," *IEEE Transactions on Pattern Analysis and Machine Intelligence*, Vol. 16, 1994, pp. 66–75.

[8] Kumar, A., "Rank Level Fusion," in *Encyclopedia of Biometrics*, S. Z. Li and A. K. Jain, (eds.), New York: Springer, 2009.

[9] Eriksson, A., and P. Wretling, "How Flexible is the Human Voice? A Case Study of Mimicry," *Proceedings of the European Conference on Speech Technology*, 1997, pp. 1043–1046.

[10] Fierrez-Aguilar, J., et al., "Kernel-Based Multimodal Biometric Verification Using Quality Signals," *Proceedings of SPIE Defense and Security Symposium*, 2004.

[11] Nandakumar, K., A. K. Jain, and S. C. Dass, *Quality-Based Score Level Fusion in Multibiometric Systems, IEEE 18th International Conference on Pattern Recognition (ICPR 2006)*, 2006, pp. 473–476.

[12] ISO/IEC, *ISO/IEC 29159-1:2010—Biometric Calibration, Augmentation and Fusion Data—Part 1: Fusion Information Format*, Geneva, Switzerland, 2010.

[13] Chang, K. I., K. W. Bowyer, and P. J. Flynn, "An Evaluation of Multimodal 2D+3D Face Biometrics," *IEEE Transactions on Pattern Analysis and Machine Intelligence*, Vol. 24, 2005, pp. 619–624.

[14] Ulery, B., et al., *Studies of Biometric Fusion*, NIST Technical Report, 2006.

[15] Snelick, R., et al., "Multimodal Biometrics: Issues in Design and Testing," *Proceedings of ICMI 2003*, 2003, pp. 68–72.

[16] Ortega-García, J., et al., "The Multiscenario Multienvironment BioSecure Multimodal Database (BMDB)," *IEEE Transactions on Pattern Analysis and Machine Intelligence*, Vol. 32, 2010, pp. 1097–1111.

[17] Phillips, P. J., et al., "The Iris Challenge Evaluation 2005," *IEEE Second International Conference on Biometrics: Theory, Applications and Systems*, Arlington, VA, 2008.

CHAPTER 12
Biometric Standards

Standards play a very important role in our everyday life. They are ingrained in virtually every product or process with which we interact. For example, the electrical plug and socket are standardized to ensure that any electronic product will work in a home. The International Organization for Standardization (ISO) defines a standard as "a document, established by consensus that provides rules, guidelines or characteristics for activities or their results" [1]. From a user's standpoint, standards ensure that products and processes perform according to an agreed predetermined expectation. From a technical standpoint, standards ensure the interoperability of products and processes. This, in turn, increases confidence of the clients and users in the technology and creates a larger market for the industry. For these reasons standards are indicative of maturity of a particular technology and industry. Although the first biometric standard was published in 1986 by NIST/ITL for use in law enforcement, virtually all biometric standards in use today were developed in the last decade. Today open standards exist for a majority of commercially available biometric technologies. This chapter will discuss the importance of standards and the standards developing organizations and provide an overview of published standards. There are several standards developing organizations, and this chapter will focus mainly on the international standards committee of ISO/IEC.

12.1 The Importance of Standards

The rapid development of biometric technologies and large-scale implementations has brought the issue of vendor lock-in, interoperability, assurance, and future compatibility to the forefront. Standards reduce the risk of depending on a single vendor by promoting interoperability and providing an alternate means of upgrading or substituting system components. This increases the confidence of end users to invest in the technology and creates a market for integrators and service providers. Integrators are a prominent force in driving the adoption of biometric technologies, and standards ensure that vendors, integrators, and end users are all on the same page. Standards generally lead to the commoditization of a technology that drives down prices, increases competition, and promotes innovation. Conformity assessment—ensuring that a product performs according to a set of guidelines—can be conducted in a consistent and repeatable manner because of standards. Data sharing is a necessity for government implementations, and standards make this possible.

12.2 Standards Development Organizations (SDO)

Standards development is carried out by two different types of organizations: formal and informal. Both of these types of organizations exist at the national and international levels. Formal organizations publish standards that are part of a legally binding contract or regulations. National standards bodies are a type of a formal organization. Generally, each country has a national body that is responsible for publishing and maintaining the standards for that country. For example, the American National Standards Institute (ANSI) is the national standards body for the United States. The International Organization for Standardization (ISO) is perhaps the most well-known international formal organization. Industry consortia and trade groups are examples of informal standards organizations. Standards created by these organizations have no legal binding, but they are normally created using a consensus-driven approach where competing views have representation and thus become heavily adopted. For example, the World Wide Web Consortium (W3C) is comprised of organizations working towards creating standards for the Web.

12.2.1 ISO/IEC

The International Organization for Standardization (ISO), along with the International Electrotechnical Commission (IEC), has jointly created a committee called the Joint Technical Committee 1 (JTC), responsible for creating information technology standards [2]. In 2002 a subcommittee SC 37 on biometrics was created within JTC 1 for developing international biometric standards. With international interest in biometrics increasing in the last decade, the membership of SC 37 has also steadily grown, with currently 28 countries registered as participating members and 11 countries registered as observers. As of the time of writing, 58 standards and technical reports had been published by SC 37. For the rest of this chapter SC 37 will refer to the international standards committee on biometrics.

SC 37 has organized its work into six working groups (WG) that bring together domain experts from all participating national bodies:

- WG1: Harmonized Biometric Vocabulary and Definitions;
- WG2: Biometric Technical Interfaces;
- WG3: Biometric Data Interchange Formats;
- WG4: Biometric Functional Architecture and Profiles;
- WG5: Biometric Testing and Reporting;
- WG6: Cross Jurisdictional and Societal Aspects.

The organizational structure of the WGs was designed to cover standardization topics from a technology and societal impact perspective. The functional relationship of the WGs ensures that standardization projects are molded by all concerned interests, whether they are technical or societal. The organizational structure of SC 37, shown in Figure 12.1, can be viewed as layers starting with technology at the base and progressively moving towards deployment concerns.

12.2 Standards Development Organizations (SDO)

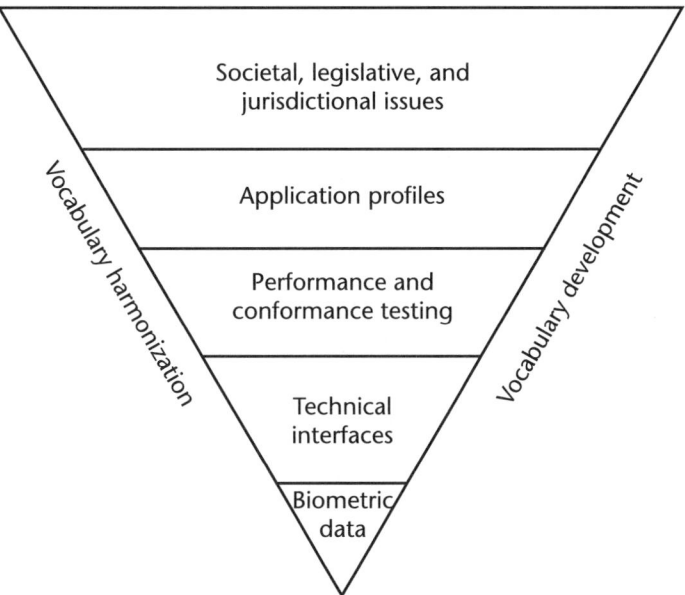

Figure 12.1 Biometric standards layers.

Standards are developed using a consensus-building process comprising several different phases, shown in Figure 12.2. The drafting process is cyclical, with participating members providing their technical input. Once consensus is reached, the draft is circulated for voting among the members, and if approval criteria are met, the draft is published as a standard. Standards have to change as technology evolves and an approved standard has to go through a reaffirmation process every 5 years. This provides an opportunity to amend and update the standard to reflect changes in technology and stakeholder concerns.

Within ISO/IEC there are other subcommittees working on biometrics as it relates to their respective areas. SC 37 liaises with these subcommittees to ensure that relevant work can be transferred and replication is avoided. They are:

- TC 68—Financial Services;
- SC 17—Smartcards;
- SC 27—IT Security.

12.2.2 National Standards Bodies

National standardization bodies are predominantly responsible for generating consensus at the national level and channeling technical contributions to SC 37. In the United States, ANSI has accredited INCITS as the primary focus for standardization of information technology. In 2001, INCITS created M1, a subcommittee for developing biometric standards in the United States. The structure of M1 mirrors that of SC 37, and M1 is responsible for establishing U.S. position and contribution to SC 37. The British Standards Institute (BSI) performs a similar role for the United Kingdom and there are several other national standardization bodies

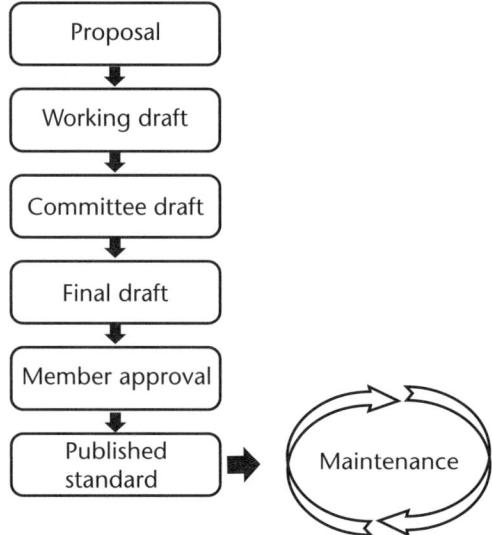

Figure 12.2 Standards development process.

around the world that provide a forum for experts to contribute to the international standardization efforts.

12.2.3 BioAPI Consortium

The BioAPI Consortium, consisting of several technology vendors, academic institutions, and government agencies, is the oldest industry group in biometrics [3]. This group was formed in 1998 to develop a vendor and technology-neutral application interface for biometric technologies. The BioAPI framework developed by this group has evolved from being an informal standard into a published standard by SC 37. This standard is discussed in Section 12.3.2.1.

12.2.4 NIST

The National Institute of Standards and Technology (NIST) is an agency of the U.S. Department of Commerce and was the first federal physical science research laboratory. NIST develops guidelines and standards for the U.S. federal IT systems for approval by the Secretary of Commerce under the Information Technology Management Reform Act. NIST develops guidelines and standards when a compelling need exists for it but is not being fulfilled by any other standards developing organization. The first biometric standard was created by the NIST Information Technology Laboratory (ITL) in 1986 for facilitating the exchange of fingerprint images between law enforcement agencies and has since played an important role in U.S. and international standards development.

12.2.5 OASIS

The Organization for the Advancement of Structured Information Standards (OASIS) is an international nonprofit that focuses on the development and adoption of open standards for information systems [4]. OASIS was founded in 1993 and currently has more than 5,000 participants from 600 different organizations and 100 countries. The experience and acceptance of OASIS as an organization for creating Web-related standards have made it a natural collaborator in creating biometrics standards for Web architectures. The OASIS XML Common Biometric Format (XCBF) Technical Committee was set up to define a common set of secure XML encodings for CBEFF patron formats, which are discussed in Section 12.3.2.1. The XCBF standard was approved in August 2003 and is freely available for download. Another technical committee within OASIS is working on defining methods and bindings by which the Biometric Identity Assurance Services (BIAS) framework can be used within XML-based Web services and service-oriented architectures.

12.2.6 International Telecommunication Union (ITU-T)

The ITU, formed in 1865, drafts and publishes international standards on information and communication technologies. ITU's Telecommunication Standardization Sector, ITU-T, made its foray into biometric standards in 2001 as part of its work on telecommunication security. The first standard, X.1081, defined a multimodal framework which can be used to classify, identify and address safety aspects of biometrics in a telebiometric system [5]. Since then, several more standards have been released, listed in Table 12.1, and there are ongoing efforts focused on different types of template protection techniques.

12.3 Working Groups

12.3.1 WG1 Harmonized Biometric Vocabulary

The field of biometrics has introduced several new concepts as well as adapted concepts from other technology areas. The objective of WG1 is to develop a harmonized vocabulary of biometric terms that already exist and develop definitions as the domain evolves. WG1 harmonizes existing terms and defines new terms using concept harmonization and term harmonization so that the ambiguity of terms is minimized. The ISO/IEC 2383-37 standard currently in draft status will be published as a standard in the near future, but currently terms and definitions are being

Table 12.1 ITU-T Standards

Name	Description
ITU-T X.1081	Multimodal model that can be used as a framework for identifying and specifying safety aspects of telebiometrics
ITU-T X.1084	General framework for authentication protocols of telebiometrics
ITU-T X.1085	Specific guidelines for authentication protocols of telebiometrics
ITU-T X.1086	Guidelines for countermeasures for improving security and privacy in telebiometrics
ITU-T X.1087	Guidelines to preserve confidentiality and integrity of biometric data
ITU-T X.1088	Framework for biometric digital key generation and protection
ITU-T X.1089	Framework for implementing a biometric system with certificate issuance, management, usage, and revocation

placed in the Standing Document 2 [6]. The document is divided into two parts where the first part outlines terms and definitions and the second part lists the relationships between existing concepts to identify gaps and terms yet to be addressed. Since SC 37 is an international organization, the degree of difficulty of translating terms into other prominently used languages is also a challenge.

12.3.2 WG2 Biometric Technical Interfaces

This WG develops standardized interfaces that allow different components of a biometric system to interact with each other. Table 12.2 lists all the standards published as a result of work done by WG2.

CBEFF and BioAPI, which are the more prominent standards drafted by this WG, are discussed next.

12.3.2.1 Common Biometric Exchange Format Framework (CBEFF)

CBEFF defines a basic data structure that supports the interchange of proprietary and standardized biometric information. The work on defining CBEFF started in 1999 when a workshop organized by NIST identified the need to create a technology agnostic data structure that could be understood by any system. The CBEFF data structure, called the Biometric Information Record (BIR), comprises three blocks:

1. The *standard biometric header* contains associated metadata regarding the biometric information. This block is mandatory.
2. The *biometric data block* (BDB) contains proprietary or standardized biometric information. This block is mandatory.

Table 12.2 Technical Interfaces

Number	Title
ISO/IEC 19784-1:2006	Biometric Application Programming Interface—Part 1: BioAPI Specification
ISO/IEC 19784-1:2006/Amd 1:2007	BioGUI Specification
ISO/IEC 19784-1:2006/Amd 2:2009	Framework-Free BioAPI
ISO/IEC 19784-1:2006/Amd 3:2010	Support for Interchange of Certificates and Security Assertions and Other Security Aspects
ISO/IEC 19784-2:2007	Biometric Application Programming Interface—Part 2: Biometric Archive Function Provider Interface
ISO/IEC 19784-4:2011	Biometric Application Programming Interface—Part 4: Biometric Sensor Function Provider Interface
ISO/IEC 19785-1:2006	Common Biometric Exchange Formats Framework—Part 1: Data Element Specification
ISO/IEC 19785-2:2006	Common Biometric Exchange Formats Framework—Part 2: Procedures for the Operation of the Biometric Registration Authority
ISO/IEC 19785-3:2007	Common Biometric Exchange Formats Framework—Part 3: Patron Format Specifications
ISO/IEC 19785-4:2010	Common Biometric Exchange Formats Framework—Part 4: Security Block Format Specifications
ISO/IEC 24708:2008	BioAPI Interworking Protocol

3. The *signature block* contains information used for security purposes such as data integrity. This block is optional.

The CBEFF BIR supports the nesting of data to accommodate multiple BDBs related to a single transaction, which eliminates the need for multiple BIRs. A CBEFF BIR is generally referred to as a patron format. A CBEFF patron is an organization that specifies one or more CBEFF patron formats in an open and public manner. Patron format A is used where a CBEFF structure needs to be used, but it is not required to establish a patron and register a format. Patron format B provides a root header for domains where CBEFF structures with multiple patron formats may be encountered. Patron format C is registered for use in BioAPI compliant applications and defines the BIR used in BioAPI. CBEFF does not specify data formats for the BDB; that work is the purview of WG3. Readers interested in CBEFF are encouraged to read the internal report published by NIST [7].

12.3.2.2 BioAPI

The BioAPI standard defines a set of interfaces at the programming level that allows multiple biometric systems to interact with each other irrespective of their underlying technology [8]. The goal of BioAPI is to reduce the complexity of developing biometric applications by abstracting technology and vendor specific details. The BioAPI architecture is comprised of the BioAPI Framework, biometric service providers (BSP), and biometric function providers (BFP). At a high level the BioAPI framework supports four basic processes: create BDB, verify BDB, identify BDB, and store BDB. The processes are performed by hardware and software that are called BioAPI units. Three types of BioAPI units are currently defined: sensor, archive, and matcher. These BioAPI units are managed by the BSP or BFP. A biometric application interfaces with the BioAPI Framework, while the BSP and BFPs interface with the vendor-specific implementation of the BioAPI unit. This architecture is illustrated in Figure 12.3.

Currently, a C programming language reference implementation is publicly available. Observing the growth of Web applications and the need to support the requirements of the developer community, a multipart standard project, ISO/IEC 30106, was initiated in 2011 to specify BioAPI for object-oriented languages. The ISO/IEC WD 30106-1 specifies the architecture for object-oriented BioAPI, and ISO/IEC WD 30106-2 defined a reference implementation of object-oriented BioAPI in Java. The goal of these standards is to create a drive for the adoption of BioAPI by allowing developers to write applications in a variety of languages, as well as take advantage of the native language API. For example, Java offers the ability to write platform and operating system agnostic applications, which is in line with the philosophy of BioAPI. Additionally, Java offers native APIs for writing applications for small-scale devices of which a BioAPI specified in Java can take advantage.

12.3.3 WG3

This WG develops standardized data interchange formats that promote the interoperability of raw and processed biometric data. This WG has the largest number of participants and is of particular relevance to biometric vendors. A multipart

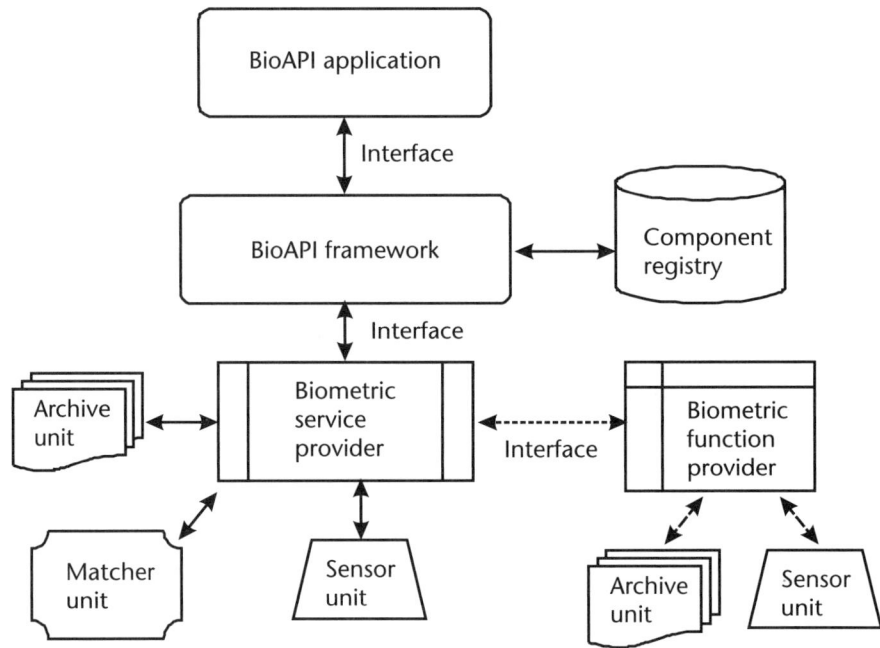

Figure 12.3 BioAPI architecture.

framework has been set up to create a standardized BDB for each biometric technology. Standards are primarily driven by vendors and end users; hence, technologies that have a higher adoption rate have been addressed first by this group. Table 12.3 lists all the completed and ongoing projects in WG3.

12.3.4 WG4

This WG develops standards to ensure interoperability of biometric information in specific implementations. Its aim is to create application profiles using base standards from other working groups, predominantly WG2 and WG3, which specify the use of optional parameters from the base standards for particular applications.

The major output of this WG is the application profiles of the multipart ISO/IEC 24712 standard. Part 2 of the standard ISO/IEC 24712-2:2008 specifies a profile that lists necessary parameters and interfaces between function modules for smart card biometric verification and identification of airport employees. Part 3 of the standard ISO/IEC 24712-3:2009 specifies a profile that defines the use of biometrics for the verification and identification of seafarers, including data storage and life cycle of the credential.

The activity of this WG is expected to increase considerably as various industry sectors such as healthcare, financial institutions, retail, and others start using biometric technologies for identity credentialing on a large scale. Table 12.4 lists the standards published by WG4.

Table 12.3 Biometric Data Interchange Format Standards

Number	Title
ISO/IEC 19794-1:2006	Part 1: Framework
ISO/IEC 19794-2:2005	Part 2: Finger Minutiae Data
ISO/IEC 19794-3:2006	Part 3: Finger Pattern Spectral Data
ISO/IEC 19794-4:2005	Part 4: Finger Image Data
ISO/IEC 19794-5:2005	Part 5: Face Image Data
ISO/IEC 19794-6:2005	Part 6: Iris Image Data
ISO/IEC 19794-7:2007	Part 7: Signature/Sign Time Series Data
ISO/IEC 19794-8:2006	Part 8: Finger Pattern Skeletal Data
ISO/IEC 19794-9:2007	Part 9: Vascular Image Data
ISO/IEC 19794-10:2007	Part 10: Hand Geometry Silhouette Data
ISO/IEC FCD 19794-11	Part 11: Signature/Sign Processed Dynamic Data (Ongoing)
ISO/IEC CD 19794-12	Part 12: Voice Data (Ongoing)
ISO/IEC CD 19794-14	Part 14: DNA Data (Ongoing)

12.3.5 WG5

This WG develops standards that define and specify biometric performance metrics, testing methodologies, and reporting requirements. Biometric testing has been researched extensively over the past two decades and the existing guidelines and best practices are extremely varied. The multipart ISO/IEC 19795 standard was established to standardize best practices for conducting various types of tests including technical, scenario, operational, usability, interoperability, and conformance tests as well as reporting the results of such tests [9]. The various parts of the standard focus on verification and identification trials with well-established ground truths, cooperative subjects, and well-defined transactions. Future work is likely to focus on complex identification applications that may work in a covert mode where establishing ground truth is a challenge.

12.3.6 WG6

The aim of WG 6 is to draft an internationally agreed upon set of principles and guidelines on how biometric technologies should be used in enterprises and governments. This WG supports the technical work of other WGs by supporting the design and deployment of biometric technologies with respect to user acceptance, policy development, privacy protection, and a common understanding of social acceptability. These nontechnical elements are essential for deployed systems because they influence user acceptance and support legal and social concerns, thus driving the adoption of biometric technologies. WG6 published the technical report ISO/IEC TR 24714-1:2008 titled *Jurisdictional and Societal Considerations for Commercial Applications—Part 1: General Guidance*, which provides guidelines for handling issues of accessibility, privacy, social concerns, and data protection [10]. At the time of this writing, a multipart standard on the use of pictograms, icons, and symbols in biometric systems was under development. ISO/IEC WD 24779-1, the first part of the technical report, provides an overview and framework for the use of pictograms and symbols to help the general public understand the use of biometric systems that can be understood irrespective of language or location of

Table 12.4 Biometric Profiles for Interoperability and Data Interchange

Name	Description
ISO/IEC 24712-1:2008	Part 1: Overview of Biometric Systems and Biometric Profiles
ISO/IEC 24712-2:2008	Part 2: Physical Access Control for Employees at Airports
ISO/IEC 24712-3:2009	Part 3: Biometrics-Based Verification And Identification of Seafarers

Table 12.5 Biometric Performance Testing and Reporting Standards

Number	Title
ISO/IEC 19795-1:2006	Part 1: Principles and Framework
ISO/IEC 19795-2:2007	Part 2: Testing Methodologies for Technology and Scenario Evaluation
ISO/IEC TR 19795-3:2007	Part 3: Modality-Specific Testing
ISO/IEC 19795-4:2008	Part 4: Interoperability Performance Testing
ISO/IEC 19795-5:2011	Part 5: Access Control Scenario and Grading Scheme
ISO/IEC 19795-7:2011	Part 7: Testing of On-Card Biometric Comparison Algorithms

the region of the world. Two other ongoing efforts, ISO/IEC WD 24779-2 and ISO/IEC WD 24779-3, provide guidelines specific to fingerprint applications and vascular applications, respectively. For nonhabituated users, sensor interaction can be a confusing experience that reduces user satisfaction and negatively impacts system performance. These pictograms and symbols, once finalized, will provide static and dynamic information to the user about the state of the system so that the user can provide the best-quality sample with the least amount of effort. For example, pictograms for fingerprint applications will convey to the user the state of the sensor, the finger that needs to be placed on the sensor, continuous feedback on the interaction to improve the quality of the captured sample, and the outcome of the capture process. The work of this WG will be crucial in the coming years as biometric systems become integrated into large ecosystems with end users from all around the world.

12.3.7 Sample Quality Standards

Biometric sample quality has a significant impact on the performance of systems, and the reliable assessment of the sample quality can lead to improved matching performance. The main challenge in quality assessment is that it is inherently a subjective process. Quality algorithms may be better aligned with certain matchers and the impact of quality could be application dependent. For example, law enforcement applications may require higher-quality samples compared to commercial applications. The multipart ISO/IEC 29794 biometric quality standard was initiated to specify the calculation and interpretation of the general biometric sample quality scores that facilitate interoperability and then extend it to specific modalities. The ISO/IEC 29794-1 standard establishes the framework for defining the meaning and data elements for the sample quality [11]. This standard does not specify the minimum quality criteria or acquisition settings for specific technologies. Two important components of the ISO/IEC 29794-1 are the concepts of normalization and calibration. Normalization is the process by which the quality score is processed to ensure a similar interpretation of the score by multiple systems. Calibration provides context to the quality score to improve predictability of matching performance. The quality score block, described in Table 12.6, is used to exchange quality information via the 19794 data interchange standards.

The other parts of ISO/IEC 29794 standard focus on specific modalities. Technical reports have been published for fingerprint and face sample quality assessment, and, at the time of this writing, a standard for the iris was being progressed through the committee. Each of these quality standards is discussed in the specific modality chapters. Table 12.7 lists work going on in the area of sample quality in SC 37.

12.3.8 Conformance Testing

Developing a standard is not enough to ensure that biometric products and systems meet the technical requirements defined in the standards. Conformance assessment is necessary to provide users with the confidence that a particular product or system will perform according to the requirements put forth in a standard. A standard provides a technical framework for making interoperability possible, while conformance assessment provides assurance that a product will perform as per the requirements of the standard. The main area of focus of conformance testing standardization is the data interchange formats specified by the ISO/IEC 19794 and ISO/IEC 19784 BioAPI standards. The ISO/IEC 29109 multipart standard establishes a generic three-level hierarchy of conformance testing methodology for biometric data interchange records as specified in the ISO/IEC 19794 multipart standard [12]. Level 1 testing checks the field values as specified in the biometric data interchange record of each base standard. Level 2 testing checks the internal consistency of the different fields and makes sure that semantic requirements are fulfilled. Level 3 testing checks that the data interchange record produced by a system is a faithful representation of the input data subject to constraints of the standard. The ongoing efforts have already established conformance testing standards for several of the ISO/IEC 19794 parts, listed in Table 12.8, while other parts are going through the drafting stages.

NIST has been involved in developing Conformance Test Suites (CTS) for various biometric standards. A Conformance Test Suite implementation evaluates the conformance to a testing methodology described in a standard. In 2006 NIST released the BioAPI CTS, which helped users establish the conformance of biometric service providers to the standard. In 2008 NIST released the CBEFF Patron-A CTS, which helped users determine whether binary file implementations conform to the standard. Currently, ongoing efforts are directed at developing a CTS for all parts of the ISO/IEC 19794 standard.

Table 12.6 Biometric Sample Quality Data Block Description

Label	Byte	Description
Quality Score	1	0: Lowest, 100: Highest, 255: Unavailable
Quality Algorithm Vendor ID	2	Registered with IBIA as a CBEFF product code
Quality Algorithm ID	2	Optional—ID registered with IBIA as a CBEFF product code

Table 12.7 List of Sample Quality Standards and Technical Reports

ISO/IEC 29794-1:2009	Part 1: Framework Biometric Sample Quality
ISO/IEC TR 29794-4:2010	Part 4: Finger Image Data (Technical Report)
ISO/IEC TR 29794-5:2010	Part 5: Face Image Data (Technical Report)
ISO/IEC NP 29794-6	Part 6: Iris Image (Ongoing)

Table 12.8 Conformance Testing Standards

Conformance testing for the biometric application programming interface (BioAPI)	
ISO/IEC 24709-1:2007	Part 1: Methods and Procedures
ISO/IEC 24709-2:2007	Part 2: Test Assertions for BioAPI Biometric Service Providers
ISO/IEC 24709-3:2011	Part 3: Test Assertions for BioAPI Frameworks
Conformance testing methodology for data interchange formats defined in ISO/IEC 19794	
ISO/IEC 29109-1:2009	Part 1: Generalized Conformance Testing Methodology
ISO/IEC 29109-2:2010	Part 2: Finger Minutiae Data
ISO/IEC 29109-4:2010	Part 4: Finger Image Data
ISO/IEC 29109-5:2011	Part 5: Face Image Data
ISO/IEC 29109-10:2010	Part 10: Hand Geometry Silhouette Data

12.3.9 Security Standards

Security techniques, specifically those related to the protection of biometric systems and biometric data, are gaining momentum within the standards community. SC 27 "Security Techniques" group within ISO/IEC JTC 1 creates standards for generic methods and techniques for IT security. SC 27 and SC 37 share a lot of common goals and a strong synergy exists in their activities. ISO/IEC 19792:2009 *Security Evaluation of Biometrics* is a standard that specifies the areas that need to be addressed during a security evaluation of a biometric system [13]. ISO/IEC 24761:2009 *Authentication Context for Biometrics (AcBio)* is a standard that defines the structure and the data elements by which a service provider can judge the results of a biometric authentication operation [14]. There are six processing units defined in this standard: capture, intermediate feature extraction, final feature extraction, storage, matching, and decision. Each of these units uses PKI to sign the data generated by them. The ISO/IEC 24745 *Biometric Template Protection* draft is currently in the standardization process and its goal is to address specific requirements of biometric information privacy such as unlinkability, irreversibility, confidentiality, and data minimization. ISO 19092:2008 is a multipart standard, developed by Technical Committee 68 Financial Services, which focuses on the secure management of biometric systems in the financial services sector [15]. Part 1 of this standard defines the security framework and part 2 specifies the cryptographic requirements for managing and securing biometric information throughout its life cycle as it is transmitted between different systems. Integrity protection of biometric data and results generated by the biometric systems are addressed in this standard.

SC 17 *Cards & Personal Identification* has worked on biometric standardization projects that are specific to card based architectures. ISO/IEC 7816:2004—Part 11 provided guidelines for the use of integrated circuit cards in personal verification using biometric technologies. ISO/IEC 24787:2010 establishes requirements for performing comparisons of biometric samples and returning decisions on an integrated circuit card and security policies for on-card biometric comparisons [16]. As biometric systems become integral components of larger infrastructures, a higher degree of collaboration is expected between various SDO committees.

A new project ISO/IEC NP 30107 titled "Anti-Spoofing and Liveness Detection Techniques" is currently progressing through SC 37 with the aim of specifying methodologies for preventing attacks on biometric sensors [17].

12.4 Standards Used in Law Enforcement

The earliest biometric standards were developed by law enforcement agencies and national governments to facilitate the exchange of fingerprint data. The ANSI/NIST ITL standard, which is the oldest standard, and its application-specific adaptations are discussed next.

12.4.1 ANSI/NIST-ITL

The ANSI/NIST-ITL standard Data Format for the Interchange of Fingerprint, Facial, and Other Biometric Information—Part 1 was drafted in 1986 as a means for exchanging fingerprint information between various law enforcement agencies and the FBI. The standard underwent revisions in 1993, 1997, 2000, and 2007. The current version of the standard ANSI/NIST-ITL 2007 incorporates data exchange for fingerprint, face, palm print, iris, scars, marks, tattoos, and future modalities [18]. A semantically equivalent standard in XML representation was published in 2008 titled ANSI/NIST-ITL 2-2008. The standard defines a set of mandatory and optional fields, including metadata, and compressed or uncompressed images. An update to this standard that will include other biometric modalities was expected in 2011. Various entities such as the Federal Bureau of Investigation (FBI), the U.S. Department of Defense (DoD), and INTERPOL have adopted this standard and specified their own set of mandatory and optional fields.

12.4.2 FBI EBTS

The FBI Electronic Biometric Transmission Specification (EBTS) is the FBI's implementation of the ANSI/NIST-ITL standard and is a successor to the Electronic Fingerprint Transmission Specification (EFTS). The FBI formats all its data using this specification and any agency that wants to match fingerprint data using the FBI system has to format its data according to FBI EBTS.

12.4.3 DoD EBTS

The DoD EBTS is used by applications requiring the exchange of biometric samples and related metadata within the DoD and other external agencies. The EBTS serves as an interface to the DoD Automated Biometric Identification System (ABIS), which contains digital and latent fingerprint information collected from persons of interest. The current version of DoD EBTS V2 was published in March 2009 [19].

12.4.4 INTERPOL INT-I

INTERPOL serves as the international law enforcement agency recognized by all its member countries. INTERPOL developed its own implementation of the ANSI/NIST-ITL standard called INT-I, which supports only the exchange of fingerprint images between member countries.

12.5 Adoption of Standards

The adoption of standards is primarily determined by the level of support by vendors and demand by end users. A standard, just like the technology it supports, undergoes an evolutionary process. A three-stage evolutionary process, along with related assessment criteria, was suggested by the National Science and Technology Council Subcommittee on Biometrics and Identity Management [20]. This process is summarized in Table 12.7.

This process has to be supported by organizations that are not necessarily involved with standards development. A formal conformance testing and certification process is necessary to validate vendor claims regarding conformance with existing standards, which in turn increases client confidence and drives adoption. NIST has established the National Voluntary Lab Accreditation Process (NVLAP), which accredits laboratories that perform conformance testing, interoperability testing, technology testing, scenario testing, and operational and usability testing for biometric systems. The *NIST Handbook 150-25* governs the technical requirements for accreditation of laboratories [21]. The overall objective of the NVLAP is to facilitate a competitive marketplace for testing biometric systems according to international standards and to increase consumer confidence in standards, which will accelerate the adoption process.

12.6 Summary

The rate of adoption of biometric technologies and its standards are closely intertwined. In the recent past, end users of biometric technologies, such as healthcare and financial institutions, have pointed to the lack of interoperability as a major barrier to adoption. To address this issue, various SDOs have focused their efforts on drafting consensus standards. The end users influence the adoption of standards by requiring vendors to conform to them. Conformance assessment standards and independent third parties to audit vendor claims will help in increasing the adoption

Table 12.7 Adoption Criteria for Standards

Level	Determination Criteria
Emerging	Standard is published by a formal SDO and is publicly available.
Stable	Standard is supported by several vendors.
Mature	Standard is implemented by several end users, and conformance testing, interoperability testing, and performance assurance testing are available from independent third parties.

Source: [19].

of biometric standards. Standards adoption can also be improved by creating central registries of approved standards and increasing awareness among vendors and end users. Standards will play a critical role in growth of biometrics in the years to come and standards development will be imperative for the mass adoption of biometric technologies.

References

[1] ISO/IEC, *ISO/IEC Guide 2:2004 Standardization and Related Activities—General Vocabulary*, Geneva, Switzerland, 2004.

[2] "ISO Technical Committees—JTC 1/SC 37—Biometrics," http://www.iso.org/iso/iso_technical_committee.html?commid=313770.

[3] BioAPI, "BioAPI Consortium," www.bioapi.org.

[4] OASIS, "OASIS: Advancing Open Standards for the Global Information Society," http://www.oasis-open.org/.

[5] ITU-T, *Recommendation X.1081—The Telebiometric Multimodal Model—A Framework for the Specification of Security Aspects of Telebiometrics*, 2004.

[6] ISO/IEC, *ISO/IEC JTC1/SC 37 Standing Document 2—Harmonized Biometric Vocabulary*.

[7] Podio, F., et al., *Common Biometric Exchange Formats Framework*, INCITS 398-2005, American National Standard for Information Technology, 2004.

[8] ISO/IEC, *ISO/IEC 19784-1:2006 Biometric Application Programming Interface—Part 1: BioAPI Specification*, Geneva, Switzerland, 2006.

[9] ISO, *ISO/IEC 19795-1: Information Technology—Biometric Performance Testing and Reporting—Part 1: Principles and Framework*, Geneva, Switzerland, 2006.

[10] ISO/IEC, *ISO/IEC TR 24714-1:2008 Jurisdictional and Societal Considerations for Commercial Applications— Part 1: General Guidance*, Geneva, Switzerland, 2008.

[11] ISO/IEC, *ISO/IEC 29794-1:2009 Information Technology—Biometric Sample Quality—Part 1: Framework*, Geneva, Switzerland, 2009.

[12] ISO/IEC, *ISO/IEC 29109-1:2009 Conformance Testing Methodology for Biometric Data Interchange Formats Defined in ISO/IEC 19794—Part 1: Generalized Conformance Testing Methodology*, Geneva, Switzerland, 2009.

[13] ISO/IEC, *ISO/IEC 19792:2009 Security Evaluation of Biometrics*, Geneva, Switzerland, 2009.

[14] ISO/IEC, *ISO/IEC 24761:2009 Authentication Context for Biometrics*, Geneva, Switzerland, 2009.

[15] ISO/IEC, *ISO 19092:2008 Biometrics—Security Framework*, Geneva, Switzerland, 2008.

[16] ISO/IEC, *ISO/IEC 24787:2010 Identification Cards—On-Card Biometric Comparison*, Geneva, Switzerland, 2010.

[17] ISO/IEC, *ISO/IEC NP 30107 Anti-Spoofing and Liveness Detection Techniques*, Geneva, Switzerland, 2011.

[18] NIST, *American National Standard for Information Systems—Data Format for the Interchange of Fingerprint Facial, & Other Biometric Information—Part 1*, 2007.

[19] Department of Defense, *Electronic Biometric Transmission Specification*, 2009.

[20] NSTC, *Supplemental Information in Support of the NSTC Policy for Enabling the Development, Adoption and Use of Biometric Standards*, 2009.

[21] Moore, B., and M. Iorga, *NIST Handbook 150-25 Biometrics Testing*, 2009.

CHAPTER 13

Biometric Testing and Evaluation Programs

Testing and evaluation are an important part of biometric systems research and development. The increase in the number of small- and large-scale biometric systems has brought more focus on the importance of credible and objective testing. Biometric system testing is unique compared to other technical systems due to the nature of the test data. Biometric system testing requires data captured from live humans; this data cannot be simulated reliably. Although synthetic biometric data generators are available, there is not enough reliable information about the distribution of features to reliably model biometric traits. This has necessitated the collection of biometric data from live users for technology and scenario testing. The results from these tests are used for making important deployment and development decisions and it is imperative that the reported performance metrics are credible. The biometrics community has organized several large-scale evaluation efforts for almost two decades to provide more clarity to implementers and end users of this technology, as well as to facilitate research by providing raw biometric data. This chapter discusses the testing and evaluation requirements, several prominent evaluation efforts, and the impact of testing and evaluation on the future of biometric systems. Readers who are not familiar with biometric system performance metrics such as FAR, FRR, EER, FMR, and FNMR should first read Chapter 2.

13.1 Biometric Testing: Why Is It Required?

Testing and evaluation are an important aspect of designing and maintaining a biometric system. The growth of the commercial market for biometrics has led to an increasing interest in ascertaining which biometric technology works the best. Comparative testing of multiple technologies using biometric data collected under similar conditions can provide this information, although it should be noted that validity of these results is applicable only to the test conditions. Various government organizations such as the National Institute of Standards and Technology (NIST) and commercial organizations have been conducting open participation tests for more than two decades. The biometrics industry is extremely fast moving and dynamic and a state-of-the-art evaluation is required to monitor its evolutionary progress. Along with assessing current technology capabilities, these tests also provide an insight into future requirements and areas of research. A majority of these tests have been conducted as independent competitions or technology evaluations based on specific operational challenges. The Fingerprint Verification Competition (FVC), conducted by an academia consortium, and Multiple Biometric Evaluation (MBE) by NIST are two prominent examples of such evaluations. The maintenance of biometric systems is an area that has received little attention compared to other deployment issues. The upkeep of a biometric system throughout its life cycle requires updating components because of technological advances or general maintenance, and these can have a significant impact on the performance of the system. An evaluation using a standardized dataset can assess the impact of the upgrade and determine if the system still functions within operational specifications.

13.2 Biometric Data Considerations

A biometric test is composed of two main parts: data collection and data analysis. The validity and reliability of a biometric test depend significantly on the type of data used, and the data collection effort has to fulfill certain criteria. The data should be representative of the target population. For large-scale applications such as voter registration, the target population is essentially the entire national population. Collecting representative data is essential, as well as being an expensive and challenging activity. Along with representativeness, the data should also reflect demographic diversity including age, gender, occupation, and ethnicity along with other relevant factors. The size of the database is closely associated with representativeness. The number of individuals and repeated samples provided by each individual should support statistical testing. The data collection activity itself is prone to human errors, especially when data is collected from hundreds of individuals. A simple mislabeling of the identity associated with the biometric record can skew the FAR and FRR results. Longitudinal data, which is collected from the same source over an extended period of time, is extremely expensive and time-consuming to collect but is more representative of real-world data. For example, multiple fingerprint samples can be collected from a subject in a session of half an hour, but data collected from the same individual spread over a period of time will allow the assessment of aging and everyday wear and tear on the fingerprint image. A real-world application will most likely not require a user to interact with the system more than three times in a single event. Best practices documents and efforts from the standards community have resulted in generally accepted data collection procedures [1–3].

13.3 Unimodal Performance Evaluations and Research Databases

The development of any new biometric technology is closely followed by a large-scale evaluation. This section will summarize some of the prominent public evaluations and research databases available for testing and evaluation. A detailed discussion of the results of public evaluations and competitions for specific modalities can be found in Chapters 3 through 11.

13.3.1 Fingerprint Recognition

The Fingerprint Verification Competition (FVC), first conducted in 2000, was conceived as an effort to determine the current capabilities of fingerprint feature extraction and matching algorithms and identifying development needs for the future. The FVC was also conducted in 2002, 2004, and 2006. The FVC is an example of a technology test that uses a common dataset to test all potential algorithms. All instances of the FVC have been open to public and have attracted participants from academia and industry. At the end of the competition, all participants are ranked based on performance metrics such as enroll rejection rate, equal error rate, false nonmatch rates at various thresholds, enrollment time, matching time, and other metrics. As the FVC has matured, the number of metrics used for the evaluations has also increased. The FVC 2004 introduced a light category for algorithms designed specifically for low memory requirements, indicating a future trend of using fingerprint recognition in embedded systems, and this methodology was also used in FVC 2006. The DET curves for open and light category on DB1 fingerprints is shown in Figure 13.1. Fingerprint data from the FVC competitions is available for researchers and can be requested from the organizers of the competition. The FVC has now evolved into a new initiative called *FVC-onGoing*. FVC-onGoing is a Web-based evaluation of fingerprint extractors and matchers on a set of four different data sets that include raw fingerprint images, challenging raw fingerprint images, 19794-2 templates, and challenging 19794-2 templates. Participants can upload their extractors and matchers and receive performance metrics on these databases. Results from the FVCs are summarized in Chapter 3 and more detailed information can be found at the FVC Web site [4].

The Fingerprint Vendor Technology Evaluation (FpVTE) was conducted in 2003 by NIST to assess the current state of the technology for small-scale and large-scale fingerprint applications. Fingerprints used for FpVTE were collected from operational systems such as visa processing and criminal booking centers. The FpVTE had three different scenarios: small-scale, medium-scale, and large-scale tests. The small-scale and medium-scale tests measured accuracy using fingerprint images from right index finger, whereas the large-scale tests measured accuracy using sets of fingerprints for an individual. A total of 34 systems from 18 participants were evaluated and the results are publicly available [5].

NIST initiated the Minutiae Interoperability Exchange (MINEX) program in 2004 with the MINEX04 test to compare the performance of proprietary template formats and ANSI/INCITS 378:2004 templates using commercial systems. The ongoing MINEX program evaluates interoperability commercial systems and certifies all systems providing false reject rates (FRR) of less than 1% at operational false accept rates (FAR) of 1%. A list of MINEX certified systems is publicly available [6]. MINEX II, which is a part of the MINEX program, evaluates match-on-card performance systems that store and match ISO/IEC 19794-2 templates on ISO/IEC 7816-compliant smart cards.

NIST has created a series of special databases that contain latent fingerprint data, scanned 10-print card data, livescan fingerprint data, and streaming video data of fingerprint captures. These databases are compressed using a variety of algorithms and are available, at a cost, for research and development purposes.

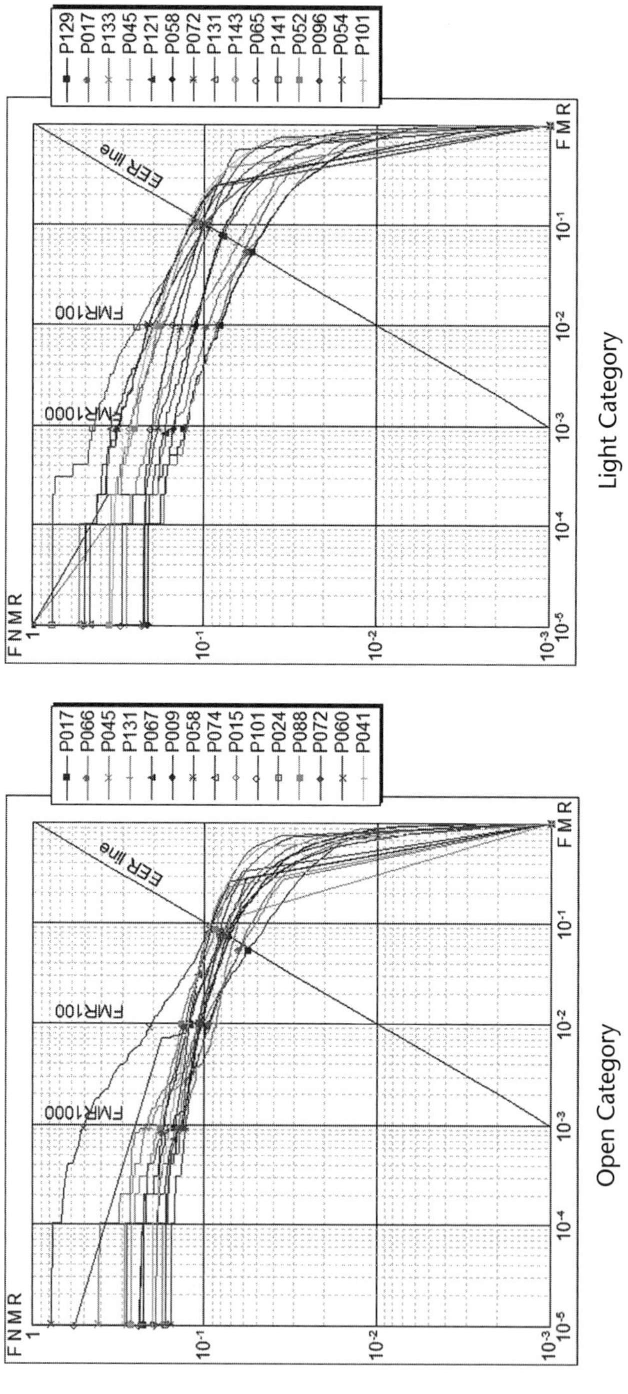

Figure 13.1 FVC 2006 DET curves on DB1 for open and light categories. (*Source:* Fingerprint Verificiation Competition, 2006. Reprinted with permission.)

These databases are described briefly in Table 13.1 and more information about them can be found on the NIST Web site [7].

The Chinese Academy of Sciences (CASIA) has collected two separate fingerprint databases, Version 1.0 and Version 5.0, which are both available for research purposes [8]. CASIA-FingerprintV5 dataset consists of 20,000 fingerprints images collected from 500 subjects on a DigitalPersona U.are.U4000 optical sensor.

13.3.2 Face Recognition

The earliest large-scale biometric technology evaluations were conducted on face recognition by NIST. The Facial Recognition Technology (FERET) program ran from 1993 to 1997. Three sequential independent evaluations were conducted using standardized tests and procedures to assess the feasibility of facial recognition

Table 13.1 Summary of Fingerprint Databases

Dataset Name	Number of Subjects	Sensors
FVC 2000	80	KeyTronic–Optical Sensor ST Microelectronics–TouchChip Identicator Technology–DF90 Synthetic Fingerprint Generator
FVC 2002	80	Identix–TouchView II Biometrika–FX2000 Precise Biometrics–100SC Synthetic Fingerprint Generator
FVC 2004	80	CrossMatch–V300 DigitalPersona–U.are.U4000 Atmel–FingerChip FCD4B14CB Synthetic Fingerprint Generator
FVC 2006	80	Electric field sensor Optical sensor Thermal sweeping sensor Synthetic Fingerprint Generator
CASIA Fingerprint Database Version 1.0	500	DigitalPersona U.are.U 4000
CASIA Fingerprint Database Version 5.0	500	DigitalPersona U.are.U 4000
BioSecure	DS2–667	Optical sensor connected to desktop
	DS3–713	Thermal sensor embedded on PDA
NIST Special Database–4	N/A	Four hundred 8-bit grayscale fingerprint pairs for five Henry classifications.
NIST Special Database–9		90 mated card pairs of segmented 8-bit grayscale fingerprint images.
NIST Special Database–14		27,000 pairs of segmented 8-bit grayscale fingerprint images and WSQ compressed; scanned at 500 dpi
NIST Special Database–24		MPEG-2 compressed digital video of live-scan fingerprint data
NIST Special Database–27		258 cases each containing latent images, matching 10-print image, and minutiae sets validated by professional examiners
NIST Special Database–29		Two hundred sixteen 10-print fingerprint card pairs with rolled and flat images scanned at 500 dpi
NIST Special Database–30		Thirty-six 10-print paired cards rolled and flat images scanned at 500 dpi and 1,000 dpi

in security and law enforcement agencies. The first FERET evaluation took place in August 1994 and was designed to evaluate algorithms on their ability to automatically locate, normalize, and identify faces from a database. This evaluation was divided into three subtests. The first subtest measured the false reject rate and the second subtest measured the false accept rate. The third subtest measured the effects of pose changes on performance. The second FERET evaluation took place in March 1995 and examined the progress of algorithms since the August 1994 evaluation. The algorithms were evaluated using a larger database and contained multiple face images of individuals collected on different days. The third FERET evaluation, which took place in September 1996, conducted a detailed analysis on the performance of algorithms under various test conditions. Four different test conditions were formulated:

1. Facial images of subjects were taken on the same day.
2. Facial images of subjects were taken on different days.
3. Facial images of subjects were taken more than a year apart.
4. Facial images of subjects were taken on the same day but with different illumination.

These four test conditions were evaluated using partially automated and fully automated algorithms. The partially automated algorithms were provided for supporting the feature detection and extraction process coordinates of the eyes. The final dataset from the three FERET evaluations consists of over 14,000 images from more than 1,200 subjects and is available to researchers under a license. Table 13.2 summarizes the data and evaluation protocols of FERET.

The Facial Recognition Vendor Tests (FRVT), another program conducted by NIST, was administered in 2000, 2002, and 2006 and performed independent evaluations of commercially available face recognition technologies. The FRVT 2000 comprised of a technology evaluation that assessed the performance of commercial systems on various face image datasets and a scenario evaluation that assessed the usability of complete face recognition systems. The technology evaluation was performed to assess the impact of compression, resolution, the distance between

Table 13.2 FERET Summary

Evaluation	Number of Enrolled Individuals	Number of Algorithms	Description
FERET 1994	316	4	Fully automated algorithms Duplicate images taken on the same day with different poses and expressions
FERET 1995	817	3	Fully automated algorithms Duplicate images taken on the same day and on different days with different poses and expressions
FERET 1996	1,196	11	Included partially automated algorithms—landmark detection performed manually Duplicate images taken on the same day, on different days, and 1 year apart Introduced an illumination variation along with the pose and expression

Source: [9].

a camera and a subject, illumination, pose, and the time lapse between the two images being compared. The results from the technology evaluation showed that significant progress had been made in the state-of-the-art systems in handling pose and illumination variations since the FERET evaluation, but the results were still not close to acceptable for real-world deployments. The product usability evaluation assessed the final distance between the subject and the camera, the ability of the system to provide a result within 10 seconds, and the ability to acquire an image within 10 seconds [10]. The FRVT 2002 was conducted to evaluate the performance of face recognition systems on large-scale datasets of images captured in real-world conditions. The test was divided into two parts: high computational intensity and medium computational intensity. In the high computational intensity test all participants had to perform identification and verification on a dataset of 121,000 full-frontal images collected in real-world conditions. The medium computational intensity test evaluated the capability of face recognition systems to compare still images under various illumination levels and pose conditions, as well as assess the impact of video data in face recognition. The FRVT 2006 specifically evaluated high-resolution still face images, 3-D face images, multisample still face images, and preprocessing pose and illumination compensation algorithms. Twenty-two participants supplied their algorithms and the results are publicly available [11]. The results from these successive tests demonstrated a reduction in FRR from 79% in 1994 FERET evaluations to 1% in 2006 FRVT evaluations.

The Face Recognition Grand Challenge (FRGC), which ran from 2004 to 2006, was established with the objective of promoting and advancing face recognition technologies designed to support U.S. government functions [12]. The starting point of the FRGC was the average performance of an 80% true verification rate obtained in the FRVT 2002 by commercially available systems on controlled, indoor face images. The FRGC set the challenge of achieving a true verification rate of 98% at an FAR of 0.1%. The FRGC was conducted as a series of six experiments with an overarching aim of examining new preprocessing techniques and developing methods for matching high-resolution images and 3-D face models to assess the best means of achieving the goal of a 98% true verification rate. A total of 19 different algorithms were submitted by various academic institutions and industry vendors from around the world, and each participant could choose to participate in one or more of the experiments. The results from FRGC showed an improvement in 2-D face recognition technology from FERET and FRVT 2000 and 2002 evaluations, and 3-D face recognition demonstrated promising results with room for improvement. The experiments with controlled 2-D face recognition yielded a true verification rate of 99.9% and the experiments with 3-D to 3-D matching yielded a true verification rate of 97%. Table 13.3 provides a high-level description of each experiment and the number of algorithms that participated in each experiment.

The Multiple Biometric Grand Challenge (MBGC) program was initiated by NIST in 2008 with the goal of addressing face and iris recognition problems that are more common in operational data. MBGC Version 1 focused on face recognition of low- to medium-resolution face images, face recognition of near-infrared illumination and high-definition video from portals, and unconstrained face recognition from still and video sequences. MBGC Version 2, which was held in 2009, collected new datasets for the same challenge areas specified by MBGC Version

Table 13.3 FRGC Description

Experiment	Number of Participants	Description
1	17	2-D images taken under controlled illumination and pose; single image used for target and query dataset
2	11	Multiple 2-D images taken under controlled illumination and pose; four images used for target and query dataset
3	10	3-D images taken under controlled conditions
3—Shape	4	3-D matching based on shape information
3—Texture	5	3-D matching based on texture information
4	12	2-D face images; target dataset consists of controlled image and query dataset consists of four uncontrolled images
5	1	Mixture of 2-D and 3-D images; target set consists of 3-D image and query images consist of controlled 2-D images
6	1	Mixture of 2-D and 3-D images; target set consists of 3-D image and query images consist of uncontrolled 2-D images

Source: [12].

1. The MBGC program was a self-selected evaluation in which participants were provided with biometric data and they voluntarily provided self-reported similarity scores for the various datasets. For MBGC Version 1, 68 organizations were provided with the data and 14 organizations submitted results to NIST. For MBGC Version 2, 78 organizations were provided with the data and 13 organizations submitted results to NIST. The MBGC has officially ended and the dataset is now available for research purposes. A description of the MGBC dataset is publicly available [13].

The Multibiometric Evaluation (MBE) program was initiated by NIST in 2009 as an open evaluation of face and iris recognition systems based on the still images and portal video sequences. The MBE still face recognition supported four different test scenarios:

1. Verification without an enrollment database;
2. Verification with an enrollment database;
3. Identification with an enrollment database containing up to 3 million records;
4. Pose calibration to assess the impact of face orientation.

The MBE still-face recognition evaluation results showed that the most accurate face recognition algorithm had a 92% chance of identifying the unknown subject at rank 1 in a database of 1.6 million records, and the identification chance decreased as the size of the database increased [14]. A subset of the identification performance graph on an enrollment dataset of 160,000 templates is shown in Figure 13.2. When human examiners are used in conjunction with the most accurate recognition algorithm, there was a 97% chance of identifying an individual when the top 50 ranked observations are returned against a database of 1.6 million records. Another interesting observation was that face recognition algorithms were more accurate on visa images compared to criminal mug shots, which can be attributed to a higher level of willingness for visa applicants to follow instructions and provide the best image possible [15].

13.3 Unimodal Performance Evaluations and Research Databases

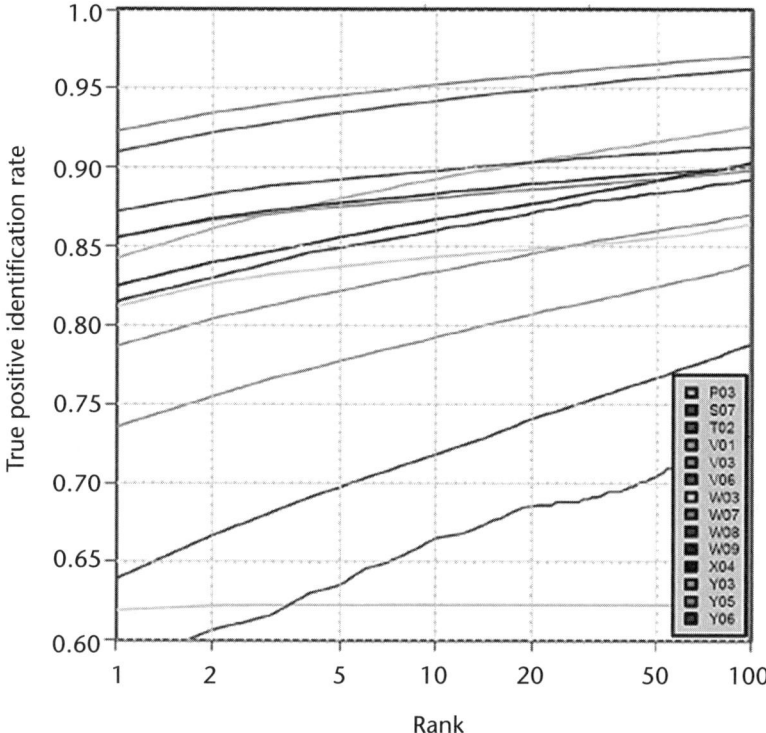

Figure 13.2 Identification performance for the MBE still-face evaluation on an enrollment dataset of 160,000. (*Source:* NIST Interagency Report 7709, NIST, 2010. Reprinted with permission.)

At the time of this writing, the MBE Portal Video Track evaluation was still ongoing. In the Portal Video Track evaluation, the goal is to recognize individuals as they walk through a portal. This supported the following two test scenarios for unimodal face recognition:

1. Match 2-D still images with high-definition video sequences.
2. Match 2-D still images in near infrared against near-infrared video sequences.

The results from the MBE Portal Video evaluation will be useful in determining the feasibility of face recognition while individuals are on the move. Such technology can be used in face recognition applications to increase throughput without interfering with the user's motion. Table 13.4 lists some of the prominent databases available for face recognition.

13.3.3 Iris Recognition

In 2005 the International Biometric Group (IBG) conducted the Independent Testing of Iris Recognition Technology (ITIRT) to evaluate the native and interoperable matching performance of three iris capture cameras using the Iridian KnoWho OEM SDK. Iridian has since been acquired by L-1 Identity Solutions, Inc. This was an independent, closed test conducted for the U.S. Department of Homeland Security to test the interoperability of iris cameras and was the first effort of its kind. This test is discussed in-depth in Chapter 5 and the results report is publicly available [16].

Table 13.4 Summary of 2-D and 3-D Face Databases

Source	Number of Subjects	Number of Images	Characteristics
Carnegie Mellon University, PIE	68	41,368	2-D images acquired under controlled levels of pose, illumination, and expression
Notre Dame University	Set B—487	33,247	2-D color images
	Set C—241	2,492	2-D infrared images
MIT	16	432	2-D images acquired under controlled levels of pose and illumination
Yale Database B	10	5,850	2-D images acquired using 585 combinations of controlled pose and illumination
Korean Face Database	Unknown	52 images per person	2-D face images of Korean subjects taken under varying poses, illuminations, and expressions
CASIA-FaceV5	500	2,500	2-D face images collected in a single session using a Logitech USB camera; variations in pose, illumination, expression and distance between subject and camera
Mobile Biometry	160	Multiple per person	Video stream database collected using mobile devices
CASIA-3DFaceV1	123	4,623	3-D scans collected using Minolta Vivid 910 camera
3D Twins Expression Challenge [15]	107 pairs of twins	428	3-D scans collected using Minolta Vivid 910 camera of twins using neutral and smiling expressions

The Iris Challenge Evaluation (ICE) was a two-phase program conducted by NIST in 2005 and 2006. The dataset for both the evaluations came from the ND-IRIS-0405 dataset, which was collected at the University of Notre Dame using an LG 2200 iris camera [17]. The LG 2200 system illuminates the ocular region using three separate LEDs. In a normal operation three images are captured sequentially and the best one is kept for further processing. This capture mechanism was modified and all three images from each acquisition were saved. Images were collected of both irises from 356 subjects, but smaller subsets with challenging data were created for ICE 2005 and 2006.

The main purpose of ICE 2005 was to evaluate right-eye and left-eye independence in terms of iris quality and recognition performance. Two different experiments were conducted for this purpose: the right iris compared to the right iris and the left iris compared to the left iris. ICE 2005 was conducted as a self-selected evaluation in which participants received the data from NIST, performed the matching operations as defined by NIST, and returned a similarity matrix to NIST. The results from ICE 2005 demonstrated that the top performers reported a true verification rate of at least 0.990 at an FAR of 0.001 and that the matching scores and quality scores for the left iris and right iris of the same individual are correlated [18]. The objective of ICE 2006, the second phase of the study, was to establish an independent performance benchmark for commercially available systems. For ICE 2006 NIST performed the matching operations on the systems that were provided by the participants. Data was collected from 240 subjects from the University of Notre Dame using the same data collection protocol but with a modified quality control module, which resulted in more challenging data compared to ICE 2005.

About one-third of the iris images were above the system-defined quality threshold and the rest were below the quality threshold. Eight groups agreed to participate in this study, but this time NIST conducted the matching operations in-house using all the algorithms. After the evaluation, only three out of the eight groups agreed to have their names and results published. The median FRR for the three systems was 0.012, 0.019, and 0.021 at an FAR of 0.001. The complete results from ICE 2006 are publicly available [11].

The Iris Exchange (IREX) program was initiated by NIST to support the development of iris recognition systems with a specific emphasis on the performance evaluation of iris image data formats, defining an iris image quality framework and performance evaluation of large-scale systems. IREX I was conducted with two specific goals in mind: (1) determining the effect of compression on performance, and (2) to gauge the ability of different vendors to match the various formats of ISO/IEC 19794-6 iris templates. The results from IREX I are available [19]. The IREX II evaluates image quality assessment algorithms and their ability to predict performance and create a standardized calibration metric for various quality algorithms. IREX II is using the Q-FIRE dataset, described in Section 13.4, collected by researchers at Clarkson University. IREX III is intended to support the development of large-scale iris systems and was scheduled to conclude by end of 2011.

The IRIS06 was funded by the U.S. National Institute of Justice and DHS and conducted by Authenti-Corp [20]. This study used three commercially available iris systems and collected data from 300 live subjects. The iris images were formatted using the ISO/IEC 19794-6 standard. This study reported acquisition error rates, FTE, FMR, FNMR, and transaction times for all three iris datasets and also concluded that left-eye and right-eye iris recognition performance is statistically similar. The lowest true match rate (TMR) and the highest TMR for three-attempt recognition were calculated to be 97.3% and 99.7%, respectively, using a threshold of a Hamming distance of 0.32.

The Soft Computing and Image Analysis Group of the University of Beira Interior in Portugal has collected iris images from subjects who cooperated minimally with the acquisition system. The goal of the database is to collect data under conditions that represent realistic user behavior and environment conditions. The first database, UBIRIS.v1, was collected in two separate sessions from 241 unique eyes with a total of 1,877 images [21]. These were still images that enforced a minimal degree of cooperation from the subject. UBIRIS.v2 is an ongoing data collection effort that at the time of this writing was comprised of 11,000 images collected from subjects on the move and in noisy conditions. Two separate public evaluations, Noisy Iris Challenge Evaluation (NICE) I and II, have been conducted using these datasets, which is discussed in Chapter 5 [20–23].

At the time of this writing, NIST was conducting the MBE with a specific scenario evaluation for comparing iris still images against still images and video sequences. Table 13.5 lists a summary of prominent datasets available for research and evaluation.

13.3.4 Speaker Verification

NIST has conducted annual evaluations of speaker recognition technology since 1996, with the latest one held in 2010. The goal of the Speaker Recognition

Table 13.5 Summary of Iris Datasets

Name	Number of Subjects	Total Number of Images	Camera	Description
CASIA 1.0 Test	1,000	10,000	IrisGuard AD100	Still iris images
CASIA 1.0	45	756	CASIA-IrisCam	Still iris images
CASIA 2.0	60 (number of unique eyes)	1,200	OKI IRISPASS-h CASIA-IrisCamV2	Still iris images
CASIA 3.0	> 700	22,034	CASIA close-up iris camera OKI IRISPASS-h	Still iris images (also includes data collected from 100 twins)
CASIA 4.0	> 1,800 real >1,000 virtual	54,601	CASIA long-range iris camera IKEMB-100 dual eye Synthetic Iris Generator	Extension of Version 3.0 dataset; three additional datasets of iris at a distance, handheld and synthetic generator
UBIRIS.v1 [21]	241	1,877	Camera Model Nikon E5700	Still iris images captured in NIR and visible light; simulates unconstrained capture conditions
UBIRIS.v2 [21]	Ongoing	11,000	Canon EOS 5D	Iris images on the move captured in NIR and visible light
ND-IRIS-0405 [15]	356	64,980	LG 2200	Three impressions of each eye captured in same presentation
University of Bath [24]	800	32,000	ISG LightWise LW-1.3-S-1394 camera, Pentax C-3516 M lens and RM-90 daylight	20 still iris images from both eyes of all subjects

Evaluation (SRE) is to assess the technical capabilities of text-independent speaker verification and provide direction for future research. Each SRE has had its own set of specific technology and scenario-related objectives that over the years have included the duration of the speech, the types of channels and microphones, pitch, ambient noise, medical conditions, and others. The main focus of each SRE is on the task of speaker detection. Although the task has remained the same over the years, additional test cases were added to each evaluation. The initial SRE focused on one-speaker detection in which the sample consists of speech from only one speaker in a conversation. Later on, two-speaker detection was introduced as a test case, in which the sample consists of a conversation between two speakers out of which only one speaker needs to be detected. The duration of the speech samples has also become a variable of interest, along with different types of microphones and landline and cellular channels.

The performance of algorithms participating in each SRE is evaluated in terms of a cost function in which the cost parameters of a missed detection and a false alarm are fixed by NIST. NIST evaluations assign a higher cost to a missed detection compared to a false alarm. The cost function described in (13.1) has been used for evaluating algorithms participating in the SRE campaign. The probability of a missed detection ($P_{Miss|Target}$) and a false alarm ($P_{FalseAlarm|Nontarget}$) is generated as part of the evaluation for each algorithm.

$$C_{Det} = C_{Miss} * P_{Miss|Target} + C_{FalseAlarm} * P_{FalseAlarm|NonTarget} * P_{NonTarget}$$
$$C_{Miss} = 10, C_{FalseAlarm} = 1, P_{Target} = 0.01, P_{NonTarget} = 0.99 \qquad (13.1)$$

Each SRE has shown a reduction in the EER from the preceding evaluation, even though the test conditions have increased in levels of difficulty. The SRE 2008 produced an EER of less than 2% for state-of-the-art systems on speaker detection tasks [25]. The performance graph for interview speech collected on different microphones is shown in Figure 13.3. At the time of this writing, the SRE 2010 results had not yet been published. The SRE 2010 uses the MIXER dataset, which contains speech data collected over landline, cell phones, and room microphones. Along with telephone conversations and interview speech recordings, the data is also categorized into high and low vocal efforts of the speaker. The evaluation is being conducted on nine different test scenarios that are testing the channel effect, effort effect, and sample length effect on the participant's algorithms. The SRE 2010 also includes for the first time the Human Assisted Speaker Recognition Test (HASR), which will be performed by systems involving human expertise to make the final decision about the trial. The level of human involvement is not specified, so systems could automate the bulk of the processing or perform minimal automated

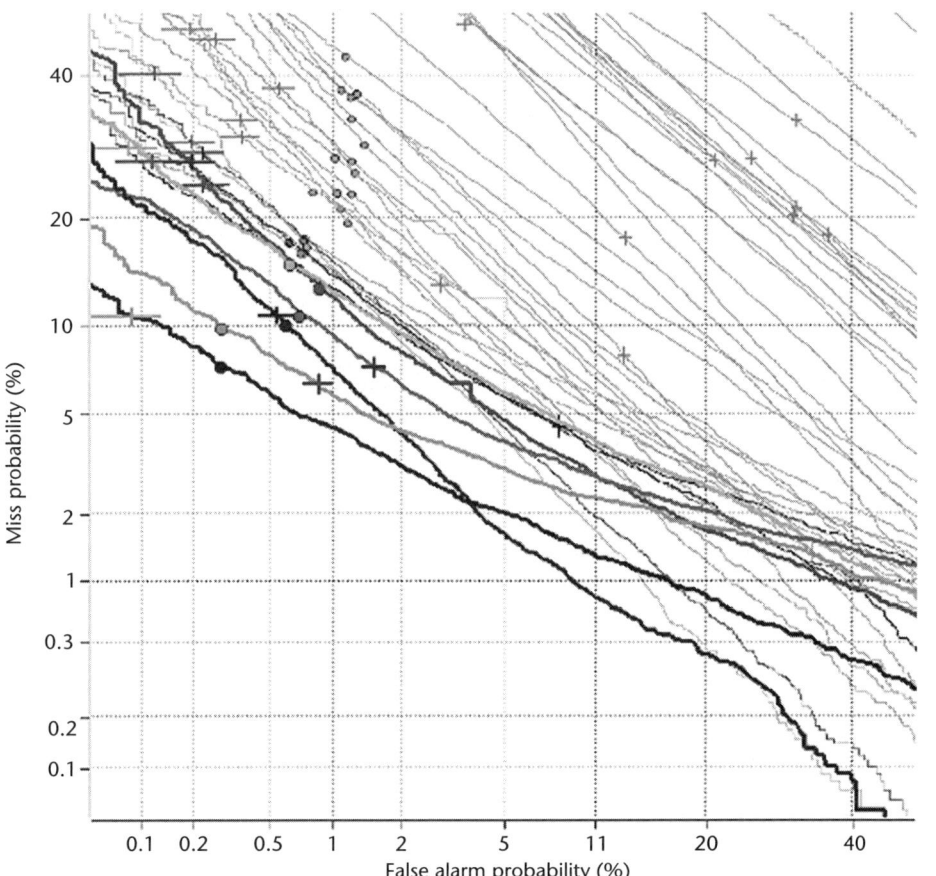

Figure 13.3 SRE 2008 performance curve for the training and testing of an interview speech on different microphones. (*Source:* The 2008 NIST Speaker Recognition Evaluation Results. Reprinted with permission.)

processing on the data. Forensic applications that require human intervention are the core audience for the results of HASR. More information regarding evaluation plans and their corresponding results is available online [26].

In 2009 IBG released the *Comparative Biometric Testing Round 7 Public Report*, which described the performance evaluation of a commercially available voice recognition system, the AGNITIO Automated Speaker Identification System [27]. Data of two time durations was used for testing: 15 seconds and 60 seconds. This data was collected over telephone and microphone channels. Identification rates were computed for all possible scenarios against an enrolled database of 500 samples. The best verification results were achieved for enrollment and verification data collected using a landline. The results for this test are discussed in Chapter 7.

Mobile Biometry (MOBIO) was initiated as part of the European Union (EU) Seventh Framework Research Programme with the aim of analyzing face and speaker recognition systems by creating a testbed of voice samples captured from mobile devices. The results of the first MOBIO evaluation were presented in 2010 [28]. The MOBIO database was collected using a mobile device at six different sites in five European countries and five separate participants provided their systems for evaluation. The data was collected from six sites in five countries in Europe. One hundred and sixty participants from across these sites completed six separate data collection sessions. In each session the subjects were asked questions and their responses were categorized into three sets: predetermined response, reading responses from an answer sheet, and free speech response. A performance metric called the *half total error rate* (HTER), which is the average of the FAR and the FRR, was used to assess performance of the systems. The results of this test were discussed in Chapter 7. A summary of speaker databases is listed in Table 13.6 [29, 30].

13.3.5 Signature Verification

The Signature Verification Competition (SVC) conducted in 2004 was the first attempt at an independent and open signature verification technology evaluation. The SVC was divided into two parts. A total of 15 international teams participated in first part and 12 international teams participated in the second part. The data collected for the SVC contained the x-coordinate, y-coordinate, time series, angle of the pen, altitude, and pressure. For part 1 only the x-coordinate, y-coordinate and time series were released to the participants. In part 2 participants could use all the captured features. The lowest and the highest EERs for part 1 were 2.87% and 28.89%, respectively, and for part 2 were 2.89% and 16.34%, respectively. The results showed that the additional information did not lead to better performance for the best algorithms, but it significantly helped the poorest performing algorithms [31].

Several large-scale signature databases have been collected with the intent of conducting technology evaluations which are listed in Table 13.7 [32].

13.3.6 Ocular Recognition

Ocular recognition is a relatively new development in biometrics that uses the face region surrounding the eye. NIST, as part of the *Face and Ocular Challenge Series*, has collected still images of the iris along with the surrounding region that are

Table 13.6 Summary of Speaker Databases

Name	Number of Subjects	Description
TIMIT	630	Read speech containing 10 phonetically diverse sentences recorded using a high-quality microphone
KING	51	Speech of 30 seconds on assigned topics recorded 10 times using narrowband and wideband telephone channels
YOHO	138	Consisted of 4 enrollment and 10 verification sessions; in each session the user had to speak a sequence of three two-digit numbers
Switchboard-1	543	Total of 2,400 two-sided conversations; conversation topics were selected from a list of 55 subjects and the conversation lasted at least 5 minutes
Switchboard-2	P1–657 P2–679 P3–640	Recording of telephone conversations between individuals from the same region of the United States, and mainly college and postcollege demographics; multiple types of landline instruments used
Switchboard Cellular	P1–254 P2–419	Recording of telephone conversations of individuals speaking on cell phones
MIXER–1 & 2	600 speakers with 10 or more calls	Recording of telephone conversations; speakers included nonnative English speakers; 200 speakers recorded conversations using four different types of channels
MIXER–3	1,867	Recording of telephone conversations of 15 or more calls; includes speech samples in 19 different languages
MIXER–4	200	Recording of telephone conversations of 10 calls; primary language of use was English, but did include different languages
MIXER–5	300	Conversation in structured interviews of 30 minutes and reading from provided text; speech recorded using 12 microphones
XM2VTS	295	Speech data recorded over 4 months
BioSecure Multimodal	Over 600	Recorded speech samples

Source: [29, 30].

available to research organizations developing ocular-based recognition algorithms. Technology evaluations and assessments are expected in this area in the near future.

13.4 Multibiometric Evaluations and Research Databases

Multibiometric systems are being deployed in large-scale applications, and several efforts have been initiated in the past decade to assess the current state and future direction of these technologies. A report titled *Studies of Biometric Fusion* was released by NIST in 2006 that presented results on a variety of score level fusion studies using fingerprints and face images of 187,000 individuals [33]. The study had two distinct objectives: (1) to assess the performance of eight different score fusion techniques, and (2) to assess the score level fusion on multimodal, multiinstance, multisample, and multimatcher architectures. Three best-performing fingerprint and face matchers from previous NIST evaluations were used in this study. Out of all the methods, multimatcher score level fusion was found to be least effective, whereas other methods that required multiple samples or instances were more effective, as shown in Figure 13.4. This indicates the positive impact of data

Table 13.7 List of Signature Datasets

Dataset	Number of Users	Attributes	Description
Philips	51	x- and y-coordinates, pen pressure, pen tilt angle	Three forgery levels: observe signing action, copy from static signature, forgery experts
BIOMET—DS1	130	x- and y-coordinates, pressure, azimuth and altitude angles of the pen	DC: Wacom Intuos2 A6 digitizer pad
BIOMET—DS2	106		DC: GripPen—Impostors forged genuine signatures using static signatures
BIOMET—DS3	91		
MCYT	330	x- and y-coordinates, pressure, azimuth and altitude angles of the pen	DC: Wacom Intuous A6 digitizer—Impostors forged genuine signatures using static signatures
BioSecure Multimodal Database—2	667	25 dynamic features extracted at point	DC: Wacom Intuos3 A6 digitizer tablet—Imposter allowed to train using video replay of signature
BioSecure Multimodal Database—3	713		DC: HP iPAQ hx2790 PDA—Imposter allowed to train using video replay of signature

*DC: data collection hardware.
Source: [32].

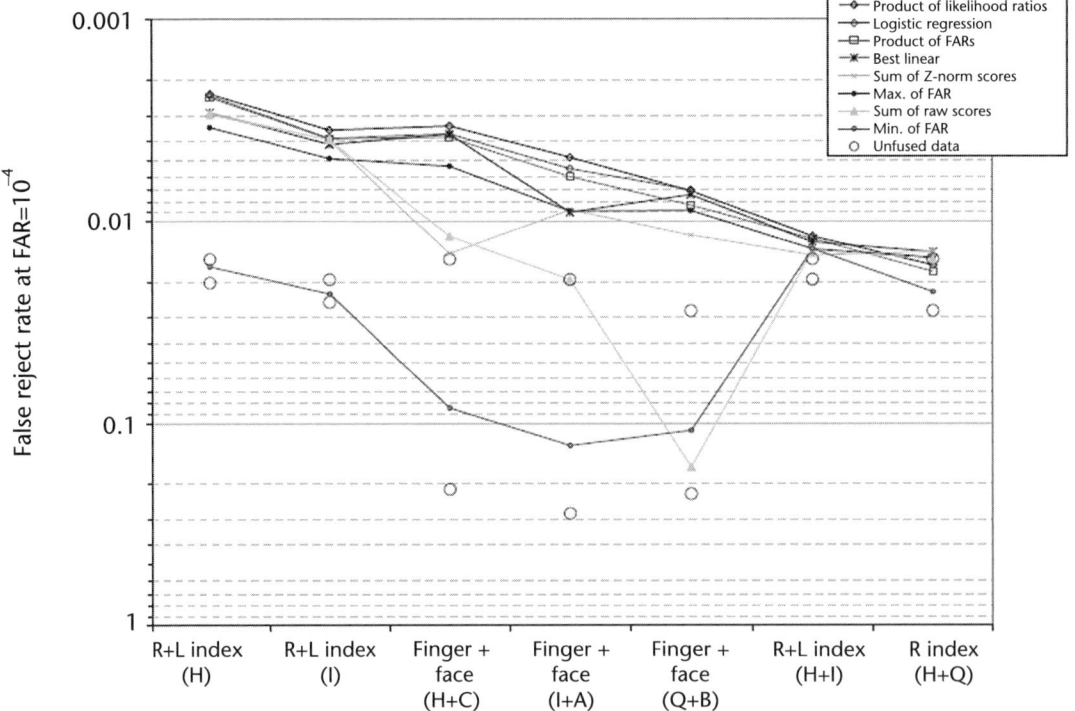

Figure 13.4 FRR (at FAR = 0.01%) for various modalities and score fusion techniques. (*Source:* NIST Interagency Report 7346, NIST, 2006. Reprinted with permission.)

independence, in the case of the finger and the face, and data variability, in the case of the multisample. The study also highlighted the need to understand the underlying probability distributions and score correlations for implementing effective score fusion methods. The results are discussed in detail in Chapter 11.

13.4 Multibiometric Evaluations and Research Databases

The BioSecure Multimodal, Multisession, and Multienvironment Database (BMDB) is one of the largest repositories of multibiometric data. More than 600 participants provided fingerprint, signature, hand, iris, face, and voice data in different scenarios. The first scenario was conducted over the Internet and collected face and audio data. The second scenario was conducted using a PC and collected fingerprint, signature, iris, hand, face, and audio data. The third scenario was conducted using a mobile device and collected fingerprint, signature, face, and audio data. This database is available to researchers through the BioSecure program [34]. BioSec and BioSecurID were two data collection efforts that were a precursor to BMDB. Several users from BioSec and BioSecurID participated in the BMDB collection. These databases were created for conducting technology and interoperability testing.

The Quality in Face and Iris Research Ensemble (Q-FIRE) dataset collected by researchers at Clarkson University comprises the face and iris videos of 195 subjects [35]. The main purpose of this dataset is to provide researchers with the capability of conducting an objective assessment of a utility of samples under various challenging capture conditions. Face and iris data were collected at various distances between 5 and 25 feet while modulating illumination levels, focus blur, motion blur, face angles, eye gaze angles, eye occlusion, and multiple faces in the capture region. The various factors were modified in a controlled manner that will allow the precise measurement of its impact on recognition performance. This dataset is currently being used in the IREX-II Iris Image Quality Calibration (IQCE) test.

Table 13.8 lists the prominent multibiometric databases available for research purposes [36, 37]. A discussion of multibiometric performance evaluation studies can be found in Chapter 11.

Table 13.8 Multibiometric Databases

Name	Number of Users	Traits Collected
BioSecure [30]	DS1–971	2-D face, voice
	DS2–667	2-D face, fingerprint, hand image, iris, signature, voice
	DS3–713	2-D face, fingerprint, signature, voice
Q-FIRE [35]	195	Video of face and iris collected between distances of 5 to 25 feet; factors affecting quality introduced into data collection protocol
MBGC	N/A	2-D face (still and video), iris (still and video)
BiosecurID [36]	400	2-D face, fingerprint, hand image, iris, keystroke, signature, voice, handwriting
BioSec	250	2-D face, fingerprint, iris, voice
MyIDEA	~104	2-D face, fingerprint, hand image, signature, voice, handwriting
BIOMET	91	2-D face, 3-D face, fingerprint, hand image, signature, voice
MCYT	330	Fingerprint, signature
BANCA	208	2-D face, voice
XM2VTS [37]	295	2-D face, voice

13.5 Comparative Tests

Comparative tests have been conducted for almost two decades. Sandia National Labs conducted an independent evaluation of signature verification, voice recognition, hand recognition, and retina recognition [38]. Error rates were calculated at various thresholds and user acceptance of the technologies was also evaluated.

Another evaluation of multiple technologies was conducted by the National Physics Laboratory in 2001. Face recognition, capacitive and optical fingerprint recognition, hand recognition, iris recognition and voice verification technologies were compared. Data was collected from just over 200 subjects on all the modalities in a normal office environment. The failure to enroll, failure to acquire, FMR, FNMR, FAR, FRR, and user throughput were calculated for each technology and the results can be found in [39].

The Mobile Biometry (MOBIO) project mentioned earlier is an example of multimodal biometric data collected using mobile devices. The performance for this evaluation was evaluated using the HTER metric, and the best performance for face recognition and speaker recognition was observed to be 10.91% and 10.59%, respectively [28]. With an increase in smartphones and devices, data of this type will be required for evaluation purposes.

IBG has been conducting Comparative Biometric Testing (CBT) since 1998 as an open performance benchmarking test of a complete biometric system. The overall objectives of the CBT are to assess matching accuracy and usability, and it combines elements of technology and scenario evaluations. The data collection is completed in real time and the results are generated offline which reflect performance of the entire system, and not just a specific subsystem. In these tests accuracy metrics measured include transactional FMR, transactional FNMR, FMR, and FNMR. Usability metrics include FTE, transactional FTA, FTA, enrollment time duration, and recognition time duration. The data collection for the CBTs happens in a controlled environment over multiple visits, and the time duration between visits and the number of visits have varied between the CBTs. The concept of transactions is discussed in Chapter 2, and in each CBT the transactions are system defined. Until now IBG has completed seven rounds of CBT, with the latest one concluding in 2009 [27]. In CBT Round 7 six commercially available systems were evaluated, but the results of only five systems are available in the public report. The systems included keystroke dynamics, iris recognition, and touchless fingerprint recognition. A total of 514 subjects participated in data collection over two separate visits in CBT Round 7, although not all of them completed both visits or provided samples on each system. During visit 1 subjects performed enrollment and recognition transactions, and during visit 2 subjects performed recognition transactions. Subjects were trained on each biometric system and assisted by test administrators during data collection. The data collection effort was tailored to the specific biometric technology. For speaker recognition, subjects provided 15-second and 60-second samples on landline and microphone channels. Two of the systems exhibited a 0.00% transactional FMR and transactional FNMR. For speaker verification more than 97% of the 15-second samples matched accurately at Rank 3. The keystroke dynamics solution exhibted an FNMR of 3.195% and an FMR of 3.264%. This test compared the behavioral and biological biometric technologies on the same subject pool, but readers should exercise caution in a direct comparison of the results. The transactions and thresholds are defined differently, which does not allow a direct comparison of performance. The data collection protocol and results for Round 7, along with Round 6 CBT, are publicly available [26, 38, 40].

13.6 Liveness Detection Evaluations

Liveness detection has started attracting the attention from the research and development community in the past 5 years. This resulted in the first Fingerprint Liveness Detection Competition in 2009 with the aim of assessing the current capabilities in fingerprint liveness detection. A universal database was created of fake and live fingerprints using three commercially available optical sensors and all participating algorithms' abilities to detect fake fingers were evaluated. The lowest overall error rate of misclassification was 14.6% and the highest error rate was 25.0% [41]. The dataset used in the competition is also available for download by contacting the organizers. Liveness detection competitions are likely to be administered for other biometric technologies in the near future. At the time of this writing, LiveDet II, a follow-up to the original competition, was being administered and results from this competition were expected towards the end of 2011 [42]. LiveDet II has introduced a new section for system-based spoof detection in which an entire system comprising a sensor and software will be provided by the participants, thereby allowing them to use hardware-based techniques along with software-based methods.

At the time of this writing, a similar competition was being conducted for the detection of face recognition spoofing attacks. The Competition on Counter Measures to 2D Facial Spoofing Attacks will evaluate the performance of algorithms in detecting printed photo attacks on face recognition systems [43]. The results were expected to be released by the end of 2011.

13.7 Summary

Testing and evaluation are necessary for the advancement of biometric technologies, and this chapter has summarized the various efforts undertaken by the biometrics community. Biometric data collection is an extremely challenging activity, and although best practices exist, future efforts are required to concentrate on standardizing the collection effort. Without such standards, the validity and generalizability of test results will be low. Another neglected area that will receive a boost from standardization is the calculation of return on investment (ROI) for biometric systems. Current ROI calculation models are still immature, and the ability to correctly quantify error rates will provide more precise information to decision makers. The results from the various performance evaluation efforts have shown a clear positive impact on furthering the state of the art and such efforts need to be applied to the latest biometric technologies.

References

[1] Wayman, J. L., and T. J. Mansfield, *Best Practices in Testing and Reporting Performance of Biometric Devices*, National Physical Laboratory, Middlesex, U.K., 2002.

[2] ISO, *ISO/IEC 19795-1: Information Technology—Biometric Performance Testing and Reporting—Part 1: Principles and Framework*, Geneva, Switzerland, 2006.

[3] ISO, *ISO/IEC 19795-2: Information Technology—Biometric Performance Testing and Reporting—Part 2: Testing Methodologies for Technology and Scenario Evaluation*, 2007, p. 48.

[4] "FVC-onGoing," 2011, http://biolab.csr.unib.it/fvcongoing/UI/Form/IJCB2011.aspx.

[5] Wilson, C., et al., *Fingerprint Vendor Technology Evaluation 2003: Summary of Results and Analysis Report*, Gaithersburg, MD, 2003.

[6] NIST, "MINEX Compliant List," 2011, http://www.nist.gov/itl/iad/ig/ominex_qpl.cfm.

[7] NIST, "NIST Standard Reference Databases: Biometrics," http://www.nist.gov/srd/biomet.cfm.

[8] CASIA, "Biometrics Ideal Test," 2010, http://biometrics.idealtest.org/findTotalDbBymode.do?mode=Fingerprint.

[9] Phillips, P., P. Rauss, and S. Der, *FERET (Face Recognition Technology) Recognition Algorithm Development and Test Report*, 1996, p. 73.

[10] Blackburn, D. M., "The Design and Implementation of the Facial Recognition Vendor Test 2000 Evaluation Methodology," M.S. thesis, Virginia Polytechnic Institute and State University, 2001.

[11] Phillips, P. J., et al., "FRVT 2006 and ICE 2006 Large-Scale Experimental Results," *IEEE Transactions on Pattern Analysis and Machine Intelligence*, Vol. 32, May 2010, pp. 831–846.

[12] Phillips, P. J., *FRGC and ICE Workshop*, 2006, http://biometrics.nist.gov/cs_links/face/frgc/PM_FRGC_Brief_FINAL_POSTED.pdf.

[13] Scallan, J. A., et al., "Overview of the Multiple Biometrics Grand Challenge," *ICB*, 2009, pp. 705–714.

[14] Grother, P. J., G. W. Quinn, and P. J. Phillips, *Report on the Evaluation of 2D Still-Image Face Recognition Algorithms*, Gaithersburg, MD, 2011.

[15] "CRVL Dataset," *Computer Vision Research Laboratory*, http://www.nd.edu/~cvrl/CVRL/Data_Sets.html.

[16] IBG, *Independent Testing of Iris Recognition Technology (ITIRT)—Final Report*, 2005.

[17] Bowyer, K. W., and P. J. Flynn, *The ND-IRIS-0405 Iris Image Dataset*, University of Notre Dame.

[18] Phillips, P. J., et al., "The Iris Challenge Evaluation 2005," *IEEE Second International Conference on Biometrics: Theory, Applications and Systems*, Arlington, VA, 2008.

[19] Grother, P., et al., *IREX I—Performance of Iris Recognition Algorithms on Standard Images*, Gaithersburg, MD, 2009.

[20] Authenti-Corp, *Iris Recognition Study 2006*, Phoenix, AZ, 2007.

[21] UBIRIS, "UBIRIS—Noisy Visible Wavelength Iris Image Databases," 2008, http://iris.di.ubi.pt.

[22] Proenca, H., and L. A. Alexandre, "The NICE.I: Noisy Iris Challenge Evaluation—Part I," *First IEEE International Conference on Biometrics: Theory, Applications, and Systems 2007 (BTAS 2007)*, Crystal City, VA: IEEE Press, 2007, pp. 1–4.

[23] SOCIA, "NICE: II Noisy, http://nice2.di.ubi.pt.

[24] Monro, D., *Smart Sensors Ltd.—Iris Capture Setup*, p. 2, http://www.bath.ac.uk/elec-eng/research/sipg/irisweb/irisimagecapturesetup.pdf.

[25] Reynolds, D. A., and W. A. Campbell, "Text-Independent Speaker Recognition," in *Handbook of Speech Processing*, J. Benesty, M. M. Sondhi, and Y. Huang, (eds.), New York: Springer, 2008, p. 779.

[26] NIST, "Speaker Recognition Evaluation," http://www.itl.nist.gov/lad/migl/tests/sre/.
[27] IBG, *Comparative Biometric Testing Round 7 Public Report*, New York, 2009.
[28] Marcel, S., et al., *On the Results of the First Mobile Biometry (MOBIO) Face and Speaker Verification Evaluation*, IDIAP Research Institute, 2010.
[29] Martin, A. F., "Speaker Databases and Evaluation," in *Encyclopedia of Biometrics*, S. Lee and A. K. Jain, (eds.), New York: Springer, 2009, p. 14.
[30] Ortega-García, J., et al., "The Multiscenario Multienvironment BioSecure Multimodal Database (BMDB)," *IEEE Transactions on Pattern Analysis and Machine Intelligence*, Vol. 32, 2010, pp. 1097–1111.
[31] Yeung, D. -Y., et al., "SVC2004: First International Signature Verification Competition," *ICBA 2004*, 2004, pp. 16–22.
[32] Garcia-Salicetti, S., et al., "Online Handwritten Signature Verification," in *Guide to Biometric Reference Systems and Performance Evaluation*, G. Dijana et al., (eds.), New York: Springer, 2009, pp. 125–186.
[33] Ulery, B., et al., *Studies of Biometric Fusion*, NIST, Gaithersburg, MD, 2006.
[34] BioSecure, "The BioSecure Multimodal Database," http://biosecure.it-sudparis.eu/AB/index.php?option=com_content&view=article&id=11&Itemid=14.
[35] Johnson, P. A., et al., "Quality in Face and Iris Research Ensemble (Q-FIRE)," *BTAS 2010*, Crystal City, VA, 2010.
[36] Orrite-Urunuela, C., et al., "Biosecurid: A Multimodal Biometric Database," *Pattern Analysis and Applications*, Vol. 13, 2009, pp. 235–246.
[37] Chan, C. H., "Extended M2VTS Database," http://www.ee.surrey.ac.uk/CVSSP/xm2vtsdb/.
[38] Holmes, J., L. Wright, and R. Maxwell, *A Performance Evaluation of Biometric Identification Devices*, Sandia National Laboratories, Albuquerque, NM, 1991.
[39] Mansfield, T., et al., *Biometric Product Testing Final Report*, Teddington, U.K., NPL, 2001.
[40] IBG, *Comparative Biometric Testing Round 6 Public Report*, New York, 2006.
[41] Marcialis, G. L., *First International Fingerprint Liveness Detection 2009*, 2009, http://prag.diee.unica.it/LivDet09.
[42] LiveDet, "LiveDet II Fingerprint Liveness Detection Competition 2011," 2011, http://people.clarkson.edu/projects/biosal/fingerprint/index.php.
[43] "Trusted Biometrics Under Spoofing Attacks," *TABULA RASA*, 2011, http://www.tabularasa-euproject.org/project.

CHAPTER 14

Desiging and Deploying Biometric Systems

Previous chapters in this book have focused on specifics of biometric technologies and related areas, but the success of a biometric implementation does not depend only on the biometric technology. Today's enterprise applications and systems are an amalgamation of several information systems, and the biometric component is only one piece, albeit a critical one, of the comprehensive infrastructure. The success of such systems depends on a holistic view of the requirements and how well the various components work together. There is a long list of failed biometric system implementations, probably none more highly publicized than the use of face recognition at the Super Bowl in 2001 [1]. Upon reviewing these systems, a list of factors that commonly contribute to the failure emerges, including:

- Incorrect selection of technology;
- Improper understanding of business or mission objectives;
- Technology capabilities overshadowing system requirements;
- Lack of understanding of the concerns of the target population;
- Inability to handle challenging users.

This chapter discusses various factors that should be considered for a successful biometric implementation. The discussion will focus on biometric-specific technology, policy, and maintenance parameters and is not meant to cover issues related to the project management of IT system implementations. There are risks and benefits of using biometrics and the goal of this chapter is to assist professional practitioners in minimizing the former and maximizing the latter.

14.1 Implementation Plan

A holistic view of the system that includes technology, administrative, and operational processes is necessary for a successful biometric system implementation. The determination of a successful implementation is subjective and a list of assessment criteria is required for the biometric implementation. A high-level list of objectives includes:

- *Flexibility*: to accommodate internal and external changes;
- *Scalability*: to accommodate the growth of the system;
- *Reliability*: to ensure consistency in the quality of service;
- *Performance*: to ensure efficient and valid decisions from the system;
- *Privacy*: to protect personally identifiable information.

Biometric system implementations need a business objective-driven approach, not just a technology-driven approach. All too often biometric system implementations are the victims of technology capabilities, and the focus on purpose is lost. A mapping of objectives to system parameters is required to guide the design and development process. Table 14.1 provides a list of factors that should be assessed and mapped to business objectives at the start of the design process.

This is a generalized list limited to biometrics-related concerns; IT system-related concerns also need to be addressed, which is outside the scope of this chapter. Cost is probably the factor that has the largest impact on any project; the discussion of cost and return on investment calculations is left to books specific to that domain.

14.2 Application Scope

The first step in the planning phase is to clearly state the objectives of identity management within the scope of the application. Identity management refers to the administration of the identity authentication process and access control privileges associated with the identity. Access control is of two types: physical and logical. The type of access control requirement will drive the entire decision process. Another consideration is if data collection will be overt or covert. Typically, all consumer-facing applications should be overt in nature, but there could be monitoring of sensitive resources that is done using covert technologies. The participation of users also needs to be decided at this stage. Certain systems may be voluntary, for example, systems that are providing a convenience to individuals such as the management of loyalty programs in supermarkets. Other systems, such as immigration and border control, are mandatory for individuals who want entry into the country.

14.3 Technology Selection

What is the best biometric technology? This is a question that is asked by all system integrators and decision makers. It is important to realize that no biometric technology can claim to be the best. Rather, the question to ask is: What is the best biometric technology *for my application*? This issue ties back to the importance given to understanding system requirements. The aspect of technology selection that gets the most attention is performance rates [i.e., failure to enroll (FTE), failure to acquire (FTA), false accept rates (FAR), and false reject rates (FRR)]. Although these parameters are extremely important, they have to be considered within the context of what is required from the application. A thorough risk analysis has to be

Table 14.1 Biometric System Implementation Design Factors

Design Factor	Specific Criteria
Application scope	Type of access control (physical/logical)
	Covert/overt
	Opt-in/mandatory
Technology selection	Type of acquisition
	Error rates, including FTA, FTE, FAR, FRR
	Deployment environment
	Target population
	Standards and interoperability
	Throughput
	Scalability
	Cultural considerations
User-system interaction	User-centric system design
Operational processes	Enrollment
	Verification
	Identification (open or closed)
	Supervised/unsupervised
	Resolution of enrollment and acquisition errors
Privacy principles	Unlinkability
	Anonymity
	Observability
	Data minimization
Architecture	Centralized storage/matching
	Specific combination of decentralized storage/matching
Application development	Integration and API selection
	Standards
Policy development	Data retention
	Scope of use
	Security of information
Design validation	Establish evaluation criteria
	Scenario testing
	Certification
Disaster recovery	Alignment with overall business continuity
	Alternative authentication mechanisms
	Recovery plan
System maintenance	Replacement of sensors
	Calibration of quality and matching thresholds
	Continuous monitoring of system health parameters
	Issue resolution for challenging users
	Hygiene procedures

performed to assess the cost of inconvenience caused to authorized users who are falsely rejected or are unable to use the biometric technology, as well as the threat exposure of allowing unauthorized users access to secure resources. An application that is striving for a high level of convenience will want to aim for a lower FRR, whereas an application that is striving for a high level of security will want to aim for a lower FAR. An implementation manager will have to sift through stacks of marketing and testing reports to assess the performance rates of different technologies. It is important to ask the technology vendors questions about how the test was conducted, how many subjects participated, what was the enrollment and verification procedure, and subject demographics to assess the generalizability of the test results to the target application. Readers are recommended to review Chapter 2 for the basics of performance assessment and how to interpret test reports.

The deployment environment has an effect on system performance, as well as on the maintenance of the biometric technology itself. If the deployment environment is controlled, then the choice of technology is quite diverse compared to a partially controlled or a completely uncontrolled environment. 2-D face recognition is affected by changes in levels of illumination, capacitive fingerprint sensors are affected by electrostatic discharge, direct sunlight on the hand geometry reader interferes with the infrared illumination and capture process, and the list goes on for every technology. The noise introduced in the biometric sample due to the environment should be minimized.

The target population is perhaps the most critical component to consider in technology selection. Age, gender, occupation, medical history, user physiology, cultural factors, and other factors significantly contribute to the performance of the technology [2]. If the demographics preclude a large proportion of the population from successfully using the technology, then it should be reconsidered. Certain cultural factors might make users disinclined towards physical interaction with the biometric sensor. The acceptance level of a technology will be driven by geographic, demographic, and cultural factors impacting the target population, and it is essential to have a deep understanding of their concerns.

Interoperability is a technology issue that is often overlooked during the design phase but becomes a critical factor during the life cycle of the application. Today's applications do not work in isolated domains; they are constantly updated and interact with multiple systems. An update to the technology should not make the application and its associated data obsolete. A technology with established and accepted standards will support the interoperability requirement and should be evaluated as part of the technology selection process.

Throughput, which refers to the number of users that a system can process in a given unit of time, is a requirement that cannot be ignored. Throughput is decided by a multitude of factors, and technology is one of them. The time taken for sample acquisition, feature extraction, and matching is technology-dependent. For example, the acquisition of both irises is quicker using a dual-iris camera than a single-iris camera, and the template generation of fingerprints is quicker than 3-D face technology.

The type of acquisition requirement between the user and the technology will also guide the selection process. The acquisition can be contact or contactless and based, and this has to be part of the technology selection process.

The ability of the technology to withstand spoofing attacks is an essential consideration, especially if the application is deployed in a nonsupervised environment. If the biometric technology is incapable of detecting spoofing attacks, it increases the risk exposure of the system and may not fulfill the underlying requirements for using biometrics in the first place.

Scalability, which refers to the ability of the technology to handle an increasing amount of information, is different for various biometric technologies. If a system is designed to handle 1,000 users today, is the underlying technology capable of handling 10,000 users in the future under the same service level requirements? Most successful systems grow in size over time and this should be considered during the technology selection process.

Cultural considerations cannot be ignored in any type of a biometric system. Concerns about stigma vary based on geocultural factors. For example, the historical use of fingerprint recognition in law enforcement carries a perception about it even when it is deployed in consumer-facing applications. If the use of a certain technology that does have an associated stigma cannot be avoided, educating users will help in dispelling such perceptions.

14.4 User-System Interaction

Sound technology design is important for a successful implementation, but a satisfying user experience is essential for a successful implementation. User interaction with the system has traditionally been a neglected aspect of system design and has contributed to the failures of several deployments. Although technology enhancements will contribute to the improvement of system performance, enhancing user-system interaction offers a greater potential for improving performance. Even though the only point of interaction between the user and the system is during sample acquisition, its impact is felt throughout the entire enrollment and recognition processes. The user's real and perceived interaction challenges have an impact on FTE, FAR, and FRR. Too often usability studies are performed after the design of the system is complete, and although this can correct some of the design flaws, the user is still treated as a functional component of the system. User-centric design makes the requirements and limitations of the end users an integral part of the system process, and the following sections will discuss the design of a user-centric biometric system. With respect to engineering design principles, user-centric design is a relatively new field, but is increasingly applied to all types of system design processes. Usability and human interaction have been identified as one of the key challenges to the mass adoption of biometrics. One of the challenges put forth by the National Biometric Challenge was to create biometric devices that are easy to use and have an intuitive interaction interface for all concerned users [3]. The Whither Biometrics Committee of the National Research Council of the National Academies observed that affordance, which refers to the notion that perception drives that action that occurs, has received very little attention in biometrics and needs to be further examined [4].

14.4.1 Usability Principles

Usability is a qualitative assessment of a user interacting with a system, product, or process. ISO has defined usability as "the extent to which a product can be used by specified users to achieve specified goals with effectiveness, efficiency, and satisfaction in a specified context of use" [5]. The usability principles of effectiveness, efficiency, and satisfaction can be viewed within the context of biometric systems as:

- *Effectiveness:* The ability of a user to consistently complete the biometric system processes as per the guidelines;
- *Efficiency:* The ability of the user to correctly complete the biometric system processes in the shortest time possible;
- *Satisfaction:* The overall experience in completing the biometric system processes.

The evaluation criteria for each principle will depend on the range of the biometric system process. For example, a system that is concerned with enrolling a user will measure effectiveness in terms of FTA and FTE, whereas a system that is concerned with only verification will measure effectiveness in terms of FAR and FRR. Supervised biometric systems have an additional factor to consider—the operator or administrator. In such a system usability from the supervisor's perspective has to be included in the evaluation criteria.

One of the earliest published works in this area examined the usability of biometric technologies for verification in bank ATMs [6]. The research examined iris recognition technology and its related usability issues for individuals using a bank ATM. By combining qualitative and quantitative techniques, a research framework was created that used focus groups to understand the user perspective of the technology, and it examined the physical constraints of using the technology, conducted laboratory testing to understand real-world issues, designed a solution, and finally conducted field trials to examine the effectiveness of the solution. A survey of clients who used the field trial unit showed that 90% were satisfied with the experience and would use iris verification over the traditional PIN and card solution.

The Visualization and Usability Group at NIST has conducted a variety of usability studies for biometric technologies used by the DHS at the port of entry. In 2007 a report on the usability testing of 10-print fingerprint capture devices was released. The US-VISIT program collected fingerprints from all 10 fingers from individuals who are not U.S. citizens or permanent residents and the objective of this study was to evaluate throughput of the biometric process. The researchers analyzed three research issues:

1. The time to capture a 10-print fingerprint image;
2. The impact of verbal, video, and poster instructional methods on user performance;
3. The most common errors and the source of the errors [7].

The results showed that subjects who had been given poster instructions had the hardest time completing the 10-print process, whereas there was no difference between verbal and video instructions. The use of operators improved the success rate of the completion process but also increased the time for capture, which is an intuitive result but had not been scientifically tested for 10-print capture systems. A related research conducted by the same group at NIST examined the height of a 10-print scanner on the capture time, the quality of captured samples, and the personal preference of users [8]. The prominent findings from the research indicated that users were satisfied with the scanner being placed at a height of 32 to 36 inches, and that for a 10-print scanner users preferred to place their thumbs separately rather than simultaneously. In 2008 the impact of the 10-print scanner angle on the capture time and the quality of samples was examined and no significant difference was observed [9]. Along with fingerprint recognition, this group has also examined usability issues with face recognition technology used by the DHS. A study in 2004 of more than 1.5 million images collected by DHS at the point of entry found that a majority of them had issues with the geometry of the image, the primary source, specifically pose, head size and position, and distortion. This research evaluated the different usability variables from the operational systems and generated usability enhancements for the face capture process. These usability enhancements were applied to a scenario test of 300 individuals and issues observed in the point of entry images were not replicated. Currently, there are several other projects being examined by this group that range from the usability of symbols for biometric technologies to anthropometric analysis of other technologies. NIST has also published a handbook on usability and biometrics that introduces concepts and design principles for improving the usability of biometric systems [10].

14.4.2 Usability Design

Usability design, as described in ISO 13407, attempts to fulfill the following objectives:

- Focus on users and environment;
- Focus on tasks to be completed by the user;
- Focus on organizational requirements and regulations;
- Integrate user concerns into system design;
- Create appropriate evaluation criteria.

The first step of the usability design process begins with gathering detailed information about the users, their tasks, the operational environment, and the functional objectives of the system. For example, the context of use of a biometric system for a hospital will be very different from that of a military facility. From a user perspective, demographic information is relevant. From an operational environment perspective, illumination, temperature, humidity, and other factors are important. From a task perspective, the importance of the task, the frequency of the task, and the impact on the organization are important. There are several methods for obtaining this information including surveys, focus groups, observational analysis, interviews, and other qualitative data collection methods. Collecting and analyzing this information helps creates personas that represent a class of users who share common traits. Certain systems might have to cater to multiple personas, and identifying this at the beginning of the design process will avoid having to integrate their requirements post hoc.

The second step involves understanding the functional requirements and nonfunctional requirements of the organization and its impact on the user-system interaction. In most consumer-facing applications, throughput tends to be a priority. In supervised applications operator functionality also becomes part of the overall requirements. A well-planned project should have collected functional requirements at the project initiation; this step should involve mapping functional requirements to design specifications.

This output should feed into the development of operational processes, applications, policy, and training to ensure a tight integration of user concerns throughout the entire development process. The usability design process is iterative and should not be conducted in a linear fashion. Every design step that incorporates usability principles should be evaluated iteratively to ensure that the most optimal design is used.

14.5 Operational Processes

A biometric system is capable of three basic processes: enrollment, verification, and identification. The key goals of all these processes is to capture the highest possible quality sample, provide a correct response in the shortest amount of time, and provide the user with a consistent and enjoyable experience. In supervised applications, the operator becomes a part of the process. An unsupervised application has a more difficult challenge of resolving errors and providing corrective recommendations in a way that the user can understand. The time elapsed between the enrollment and verification or identification attempt is another factor to consider. Certain biometric templates, such as hand recognition templates, change to a higher degree compared to fingerprint templates. Either template adaptation or the requirement for periodic verification or identification attempts should be incorporated into the operational processes. Acquisition errors are bound to occur in these processes; the number of successive errors allowed will impact the throughput rate as well as the performance error rates. A certain percentage of the target population will not be able to enroll in the biometric system. In such a case an alternative authentication process is required. In the case of an error during verification or identification, there should be an adjudication process to finalize the outcome. The type of identification—positive or negative, open or closed—needs to be addressed. The biometric system-specific processes have to integrate with the rest of the application to provide the user with a seamless experience.

As part of the user-centric design, the operational processes should also take into account the context of use. Information regarding independent and dependent tasks, the initiation and termination of each task and subtask, and the type of feedback given to the user needs to be incorporated into the operational process. For example, if an iris recognition system is triggered only if the user stands at a certain distance from the camera, is this information clearly understood by the user? A solution would be to draw a rectangular box on the ground that indicates to the user where he or she should stand in order to initiate iris capture. These operational processes represent interaction between the system and user that needs to be designed, keeping effectiveness, efficiency, and satisfaction in mind.

14.6 Privacy Principles

The advantage of biometric technologies arises from a strong binding of a user with physiological or behavioral traits, but this also raises several privacy concerns. Biometric information is immutable and can act as an identifier to link disparate databases, and the threat of function creep in government-run systems increases the probability of its misuse. Biometric information qualifies as personally identifiable information (PII) and has to be treated accordingly. Privacy protection has to be addressed as a design principle, both from technical and policy perspectives, within the context of system's objectives. An example is the ISO 19092:2008 standard, prepared by Technical Committee 68 Financial Services, which is focused on the secure management of biometric systems in the financial services sector. This standard defines requirements for managing and securing biometric information throughout its life cycle and is implicitly linked to addressing the privacy concerns of users. Chapter 16 discusses in depth the various privacy concerns that should be addressed as part of a larger system.

14.7 Architecture

There are several possible combinations when it comes to architecture of the biometric components, which is largely driven by data storage and matching locations. The storage and matching locations can be categorized into four general groups: local machine, cards/tokens, devices, and centralized server [11]. A matrix showing all possible combinations is illustrated in Chapter 1 (Figure 1.8), but it should be noted that some of the combinations are not practical.

The type of architecture will depend on the scope of the application. Large-scale identity management systems like border control and national ID programs will typically go for a centralized storage/centralized matching or card storage/centralized matching type of architectures. A time and attendance application can be designed as a device storage/device matching architecture. The number of simultaneous users, geographic locations of the authentication service, and security threats are also parameters to consider in architecture design.

An increasing number of implementations are now moving towards card-based architecture. Privacy and security concerns are a major driver for card-based architectures, and three different types have emerged in the last decade. *Template-on-card* architecture stores the enrollment template on a smart card or as a barcode and uses it during the matching process. *Match-on-card* architecture uses a smart card that is capable of storing and matching biometric samples. The data acquisition is done using a peripheral device that is capable of communicating with the smart card. *System-on-card* architecture is a self-contained system that includes a sensor, feature extractor, storage, and matcher on a card or token. The security of all these architectures is based completely on the tamper resistance of the card and the security of the communication channel between the card and the enterprise application.

14.8 Application Development

Application development requires a thorough understanding of functional requirements of the system as well as operational and business processes. The application development process can be divided into two stages: the development of the biometric application and the integration with nonbiometric components. Once the technology and the processes have been decided and the architecture has been designed, the next step is to develop the biometric application based on these specifications. The choice of programming language, the application programming interfaces (API), and the deployment environment are important parameters to consider in application development. The BioAPI, which defines a standardized programming interface that is technology and vendor neutral, is an example of API selection [12]. The BioAPI reduces the application development time and reduces the binding of the application to a specific technology. This is especially useful when a specific biometric component such as the sensor, feature extractor, or feature matcher is upgraded or replaced since the entire application does not have to be reengineered.

The biometric system is likely to interact with application servers that support the business operations, data backup systems, PKI, user directories, and other information systems. Network configuration and communication between various systems have to be handled in an efficient and secure manner. Application integration introduces the most number of security vulnerabilities and operational bottlenecks. These issues can be addressed by creating a cross-functional application development team that includes members who are experts in the underlying technology, network configuration, databases, and business processes.

14.8.1 Application User Interface

The application is the primary means of providing feedback to the user or the operator in a supervised system. The means of communication should factor in user demographics and the operational environment. For example, a face recognition system could employ auditory feedback or visual feedback mechanisms. If the face system is going to be used by individuals of various nationalities who do not share the same primary language, the auditory feedback mechanism will not be effective. Instead, a visual feedback system that gives the information in a graphical format would be more effective, efficient, and satisfying.

14.9 Policy Development

Policies are important to support and maximize the potential of the underlying technology. The existence of policies provide users with the confidence that the management of the system has been given due diligence. The question at the top of most users' minds is: What is going to happen with my biometric information? A formal policy or list of policies outlining the following is required:

- Scope of application;
- Scope of use of the biometric information;
- Time of retention of the biometric information;
- Security controls on the data;
- List of individuals who have access to the biometric information;
- Data-sharing agreements.

As the scope of biometric systems expands and user awareness increases, policies will take on an increasingly important role. The Biometrics Institute in Australia published the *Biometric Privacy Code* report with guidelines on how biometric data should be collected, used, and disclosed to protect privacy of users [13]. The report addressed three specific goals:

1. To facilitate the protection of personal information provided by or held by biometric systems;

2. To facilitate the process of identity authentication in a manner consistent with the Privacy Act;
3. To promote biometrics as privacy-enhancing technologies (PET).

Organizations will have to balance their own requirements with best practices to create policies that address user concerns.

A privacy impact assessment should also be performed as part of the planning process. This process will help in assessing and identifying privacy concerns and addressing them through the formulation of policies, as well as any architectural redesign. Chapter 16 explains this framework in the biometrics context in more detail.

14.10 Design Validation

Design validation is an essential part of system design to ensure that the system performs according to expectations and fulfills the functional objectives. Design validation can be categorized into *data validation*, *requirements validation*, and *functional validation*. Data validation is done to ensure that the input and output of the subsystems are semantically and syntactically fulfilling design requirements. The use of standardized data interchange formats can be validated using conformance test suites (CTS) discussed in Chapter 12. Requirements validation will ensure that the various components of the system adhering to the demands identified during the requirements analysis phase are met. This can include the conformance tests of standardized profiles and interfaces, as well as legal and regulatory audits. Depending on the scope of the system, functional validation should include one or all of the following: laboratory testing, environmental testing, and scenario testing with the target population. The various performance metrics, such as FTA, FTE, FAR, FRR, and throughput established to measure functional success, should be applied to a pilot of the system. It is preferable that the design validation is conducted by individuals who were not an integral part of the design and integration process.

14.11 Disaster Recovery Plan

All systems are hit by unforeseen disasters, and it is imperative to have a plan to continue with business operations while the system is recovering. These plans focus on bridging the gap between software, hardware, and data that has been irrevocably damaged while the system is restored to normal operations. If identity management is a necessary component to business operations, a disaster recovery plan is essential. This plan should outline alternative authentication methods, the restoration of biometric information from backups, the installation of new biometric devices, the recapture of biometric information if essential, modified verification and identification processes, and the documentation of system restoration. This disaster recovery plan should be integrated into the business continuity plan of the organization.

14.12 Maintenance

The success of a biometric system cannot be ascertained based on successful design and deployment. Management and maintenance of the system throughout its life cycle will contribute significantly to its overall success. A biometric system is no different than any other network or system; it also requires an administrator who is responsible for handling the day-to-day operations, which include the maintenance of the biometric sensors, replacing nonfunctional sensors, adding and removing users, and resolving other technology and user-related issues. A critical aspect of ongoing management is the performance monitoring of the biometric system. All biometric systems have a recommended quality and match threshold associated with it. The threshold could require recalibration based on a review of the operational performance or a change in the operational requirements of the system. Real-time performance monitoring can help in identifying throughput bottlenecks and challenging users. A subset of the enrolled users could be responsible for a disproportionately large number of the errors. A periodic zoo analysis of system users will help in identifying such users and correct the user-specific issues. Ongoing management, if done correctly, will ensure that the system achieves optimal performance.

Biometric sensors undergo a lot of wear and tear, especially when they are placed in an open environment and are for public use. The comfort level of the users with the system correlates with how hygienic they perceive the system to be, which also holds true for most publicly used resources. Biometric sensors that require physical interaction, such as fingerprint and hand recognition, create a greater hygiene concern for users. A fingerprint sensor that looks dirty or a hand recognition device with residue on it is likely to make users question its hygiene safety. A procedure to maintain a certain standard of hygiene is an essential part of regular maintenance. However, this procedure should take into account the impact of cleaning the sensor based on its underlying technology as well as the impact of the cleaning agent on the user's biometric characteristics. For example, cleaning capacitive fingerprint sensors with chemical cleaning agents corrodes the sensing surface and reduces its efficacy. Providing users with alcohol-based sanitizers is another preferred means of maintaining hygiene standards where any type of physical interaction is required, but alcohol-based sanitizers reduce the moisture content of the skin, which can have an adverse impact on fingerprints captured using an optical sensor.

Auditing the biometric component is a prudent step to take as part of maintenance. It is better if the auditing is conducted by an independent authority, but an internal team can also be used. The Information Systems Audit and Control Association (ISACA) has released a document describing audit guidelines for information systems that use biometric technologies [14]. The guidelines are developed for auditors to ensure that business objectives are being met by technology and operational processes. The audit process is presented as a six-step process:

1. Selecting and acquiring biometric system;
2. Operation and maintenance of biometric system;
3. User training and acceptance;
4. System performance;

14.13 Summary

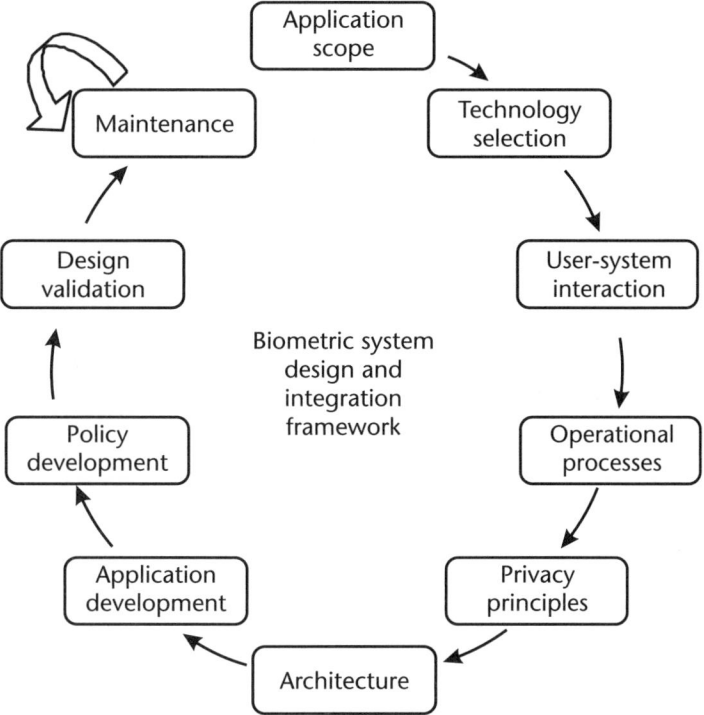

Figure 14.1 Biometric system implementation framework.

5. Application and database controls;
6. Audit trails.

14.13 Summary

Figure 14.1 summarizes the implementation framework that should be used for biometric systems. There are many small- and large-scale biometric systems deployed today that are being used in a wide variety of applications. The majority of these applications interface with internal and external systems, and new deployments will be even more tightly integrated. Successful biometric implementations are driven by a mix of factors, including project management and requirements analysis, technology, policies, and maintenance of the system. This chapter briefly discussed the factors that have an impact on systems that use biometric technologies and provided designers and implementers with a road map towards a successful deployment. By taking into account administrative, technical, and organizational goals and making them an integral part of the design and development process, biometric systems can be made a more integral part of the IT infrastructure.

References

[1] McCullagh, D., "Call It Super Bowl Face Scan I," *Wired.com*, 2002.

[2]   Wayman, J. L., and T. J. Mansfield, *Best Practices in Testing and Reporting Performance of Biometric Devices*, National Physical Laboratory, Middlesex, U.K., 2002.

[3]   National Science and Technology Council Subcommittee on Biometrics, *The National Biometrics Challenge,* August 2006, p. 22.

[4]   Pato, J. N., and L. I. Millett, (eds.), *Biometric Recognition Challenges and Opportunities*, Washington, D.C.: National Research Council, 2010.

[5]   ISO, *ISO 13407:1999 Human-Centred Design Processes for Interactive Systems*, 1999.

[6]   Coventry, L., A. De Angeli, and G. Johnson, *Usability and Biometric Verification at the ATM Interface*, New York: ACM Press, 2003.

[7]   Theofanos, M., et al., *Usability Testing of Ten-Print Fingerprint Capture,* National Institute of Standards and Technology, 2007, p. 56.

[8]   Theofanos, M., et al., *Effects of Scanner Height on Fingerprint Capture,* National Institute of Standards and Technology, 2006, p. 58.

[9]   Theofanos, M., et al., "Does the Angle of a Fingerprint Scanner Affect User Performance?" *Human Factors and Ergonomics Society Annual Meeting Proceedings*, Vol. 52, 2008, pp. 1989–1993.

[10] Theofanos, M., B. Stanton, and C. A. Wolfson, *Usability & Biometrics: Ensuring Successful Biometric Systems*, NIST, 2008.

[11] INCITS M1.4, *Study Report on Biometrics in E-Authentication*, 2007.

[12] ISO/IEC, *ISO/IEC 19784-1:2006 Biometric Application Programming Interface—Part 1: BioAPI Specification*, 2006.

[13] *Biometrics Institute Privacy Code*, Biometrics Institute, Crows Nest, NSW, Australia, 2009.

[14] ISACA, *G36 Biometric Controls*, Rolling Meadows, IL, 2007.

CHAPTER 15

Biometric System Security

A system that provides security functionality can itself become a target of threats. This is not a new phenomenon; cracking password programs and getting past firewalls are the digital equivalent of picking locks. Security means different things to different people. However, the overall objective of security remains unchanged: to prevent individuals from circumventing controls that protect a critical physical or digital resource. In the context of biometrics as well, security has several connotations. Other chapters in this book have discussed issues related to extraction and matching errors and the privacy of sensitive information. This chapter will discuss security from a system perspective and mainly focus on security challenges of the data acquisition stage since that is the only point of the system that is publicly exposed. Some of the commonly expressed security issues with biometric systems and techniques for vulnerability exploitation and risk mitigation are also covered.

15.1 System Security Analysis

A biometric system needs to offer its end users a certain level of trust in the authentication process and in maintaining the privacy of their sensitive information. Raw biometric data is immutable and, in some cases such as face recognition, not confidential. Given these characteristics, there is a significant payoff for attackers to circumvent a biometric system. As the number of system deployments increases, such deployments are bound to face newer and more innovative attacks. A thorough security analysis of a biometric system should include the examination of vulnerabilities of each subsystem, the interaction between each subsystem, and the operational processes. This is an active area of research and there exists a substantial amount of literature that discusses various attack vectors. The initial work by Ratha et al. identified eight different attack points in the general biometric model [1]. Subsequent research based on the attack tree model identified 20 attack points and 22 different vulnerabilities [2]. This analysis framework included administrative processes, the enterprise IT system, and token presentation. Figure 15.1 illustrates the different vulnerabilities that can potentially be exploited in a biometric system.

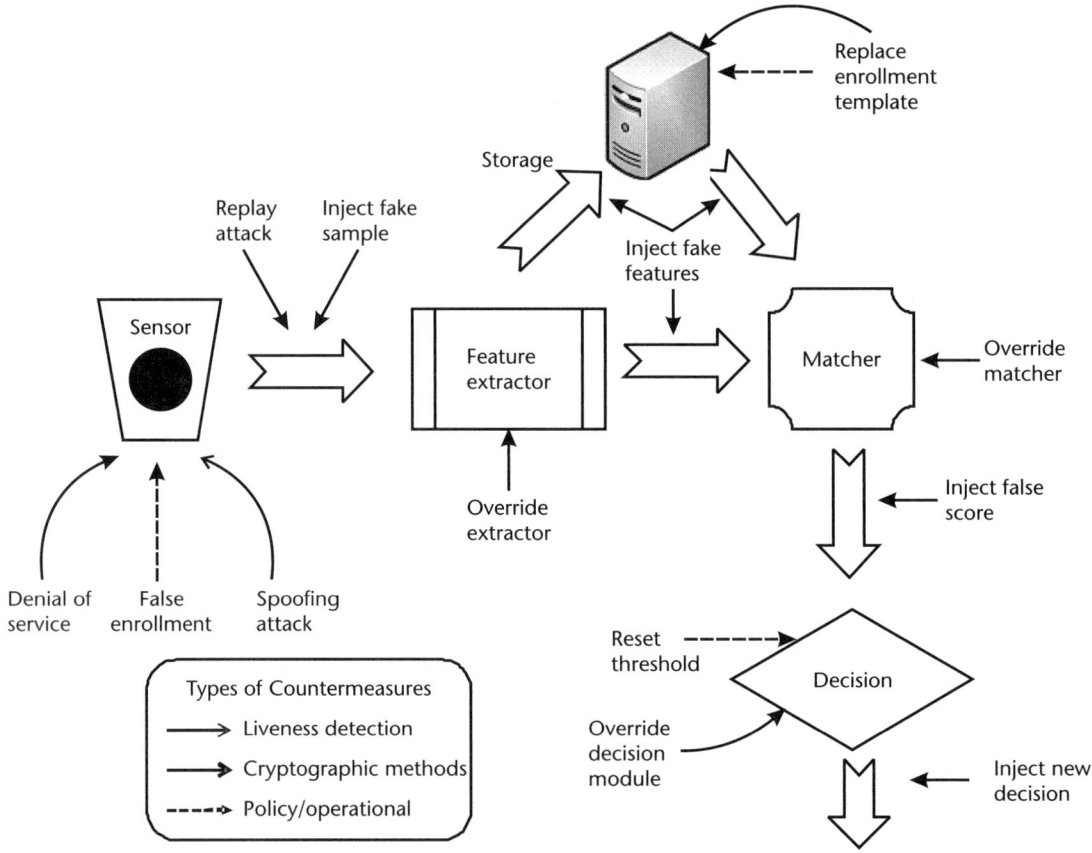

Figure 15.1 Security analysis of biometric system and processes.

15.1.1 Subsystem Vulnerabilities

The acquisition system is the sole point of contact between the user and system, which also exposes it to several threats. The acquisition system can be presented with fake biometrics to impersonate an individual who is already enrolled in the system. Such an attack is called *spoofing*, and it is discussed in Section 15.2. In negative identification applications such as watch lists, individuals who want to avoid detection will provide samples that do not match the enrolled template. For example, smudged fingerprints or a significant change in facial hair can deceive some fingerprint and face recognition systems, respectively. Hill-climbing attacks on biometric systems use prior knowledge of match scores to recreate a biometric sample through multiple iterations [3]. Systems that require physical contact with the sensor often leave behind residue in the physical shape of the biometric trait. Optical touch fingerprint sensors are a classic example of this phenomenon. Attackers can use the residue to create fake biometric samples or fool the sensor into believing that the residue is a live sample. Sometimes attackers might be interested only in denying service or access to resources and not in gaining access to a system. Such attacks are called denial of service (DoS) attacks and can be directed at acquisition sensors by breaking or affecting the acquisition ability of the sensor.

The signal processing subsystem segments the sample and extracts unique features. Replacing this subsystem so that it produces only a certain type of feature vector can subvert the template creation and matching process. A hill-climbing attack can be initiated by targeting this subsystem, with the condition that a matcher is providing similarity scores. A successful DoS attack can be launched by preventing feature extraction and thus increasing the FTE and FTA. Quality control is also a function of this subsystem. Subverting the quality control module will allow an extremely poor sample to be saved as an enrollment template. Low-quality samples increase the probability of false accepts and the probability of success or an impersonator.

The data storage subsystem is vulnerable to data theft and replacement attacks. By replacing the biometric template of an unauthorized individual, an attacker can gain access to the system using his or her own sample. If raw biometric samples are stolen from the data storage, they can be used in transmission attacks, which are described in the next section. Although it is generally considered that templates cannot be used to regenerate fingerprint images, researchers have demonstrated the ability to synthetically generate a fingerprint image using minutiae points from standardized ISO/IEC 19794-2 templates [4].

The matching subsystem compares two biometric samples and provides a similarity score. An attack based on zoo analysis can be executed if access to the matching subsystem is available. Wolves and lambs result in a large number of false accepts in biometric systems. This information can be used to find individuals who might be wolves or lambs for a particular system and use them as impersonators. Overriding the matching subsystem so that it always provides a predefined similarity score is another type of attack on this subsystem.

The decision subsystem provides the final verdict on pass/fail of the identification or verification process. A basic attack targeting the decision subsystem will always provide a predefined decision that fulfills the objective of the attacker. The matching threshold of the decision subsystem can be modified to favor a certain type of error that will launch either a DoS or a masquerade attack. The attacks on different subsystems are summarized in Table 15.1.

15.1.2 Transmission Vulnerabilities

An increasing number of biometric systems today are networked with external information systems such as healthcare record management systems and manufacturing control systems. Data that is transmitted using wired or wireless means is vulnerable to data snooping, man-in-the-middle, and replay attacks. In these attacks the attacker obtains access to the data transmission between systems and then exploits it for malicious purposes by either replacing the authentic data or injecting data captured from a previous transmission. For example, fingerprint information transmitted to the database for storage can be intercepted and then used to initiate another transaction under the identity of the authentic user. Most biometric devices connect to a computer using Universal Serial Bus (USB) ports. There are several hardware and software methods available in the public domain for recording all data exchanged between a hardware device and the device driver. An unattended system can be attacked to collect this data from authentic users and then used to exploit the system. Such attacks are not specific to biometric systems; they affect all

Table 15.1 Security Threats and Mitigation Strategy

Category	Threat	Mitigation Strategy
Acquisition	Spoofing/fake biometrics	Liveness detection
	Hill climbing	Providing low granularity response
	Inserting fraudulent sensor	Check for device ID
	Using residue	Cleaning sensor/liveness detection
Signal processing	Replacement attack	Integrity check/hashing
	Hill climbing	Providing low granularity response
	DoS	Monitoring performance of system
	Quality control	Integrity check/hashing
		Monitoring performance of system
Data storage	Injecting unauthorized templates	Database security
	Stealing authentic templates	Encryption
Matching	Inserting fraudulent subsystem	Integrity check/hashing
Decision	Inserting fraudulent subsystem	Integrity check/hashing
Transmission	Replay/man-in-the-middle	Encryption, digital signatures, time stamps
Operational process	Enrollment (document vetting)	Creating document-vetting policies and procedures
	Verification/identification (hill climbing)	Allowing a predefined number of successive failed attempts

systems that are internetworked and transmit data. There exists a wide variety of solutions based on encryption and hashing algorithms that have been extensively tested for preserving confidentiality and integrity, and a few have been standardized. A discussion of these solutions is beyond the scope of this book; interested readers will find detailed information in any computer and network security book.

15.1.3 Process Vulnerabilities

The performance of a biometric system is impacted by the enrollment, verification, and identification process; the same is true of system security. The enrollment process is of the utmost concern since it determines the identity that is linked with the biometric sample as well as the access privileges. Inadequate vetting of the ground truth credentials such as a driver's license or a passport can lead to duplicate or fraudulent identities being enrolled in the database. During verification or identification, the fraudulent identity can be used for recognition without any issues. Hill-climbing attacks, discussed in Section 15.1.1, are difficult to execute in real-world systems but not impossible. Systems can be protected against such attacks by providing pass/fail or coarse responses instead of finely granular responses or precise match scores or only allowing only a certain number of fail responses before alerting the administrator. In several systems biometric technologies are used for the rights and access management to critical resources. Unauthorized modification of access privileges can result in individuals accessing information beyond their privilege level.

15.2 Spoofing and Mimicry Attacks

In spoofing attacks a fabricated biometric sample of a physical trait is used to deceive the system into believing that the sample is provided by a live and authentic user. Mimicry attacks replicate behavior of the genuine user to deceive behavioral biometric systems. Such attacks are not new; they have been carried out from the time that fingerprint recognition became a standard for law enforcement. In 1920s Alert Wehde used his photography and carving skills to generate a fake fingerprint surface that could be used to leave behind fake latent prints at the crime scene [5]. Since fingerprint recognition is the most widely used biometric technology, it is also the target of most spoofing attacks. Spoofing attacks on face recognition and speaker verification followed very soon after they were shown to be feasible. The development of other biometric technologies has led to newer and more innovative attacks. The following sections will discuss spoofing attacks and mitigation strategies for various biometric technologies.

15.2.1 Fingerprint Recognition

The majority of fingerprint recognition systems use minutiae points for comparing two fingerprints. Most live capture systems require a finger to physically interact with the device surface for it to capture the fingerprints. In touch type fingerprint sensors, this leaves behind residue that form latent fingerprints. Optical sensors use the difference of reflectivity between the ridges and valleys to capture the fingerprint image. If the latent fingerprints are thick enough, the optical sensor could detect it as a real fingerprint image. Researchers have also shown that low end optical and capacitive sensors can be activated by breathing on the platen of the sensors as it increases the visibility of the latent prints [6]. The same research group successfully spoofed an optical fingerprint sensor by placing a thin walled plastic bag filled with water on the platen. A technique that uses a printout of the fingerprint pattern with embossed ridgelines was also successful in spoofing the system [7]. Forensic methods for lifting fingerprints off surfaces can provide a highly detailed fingerprint structure depending on the amount of residue. Researchers transferred this latent print onto a transparent adhesive tape that was then placed on a fingerprint sensor and successfully recognized.

There are several methods that use physical molds for creating 3-D fingers with fingerprint patterns printed on the ventral side. In 2000 a research paper was published that showed how to deceive fingerprint sensors using silicone rubber [8]. Fingerprints were lifted from cooperative and noncooperative users and then used to successfully spoof four optical and two capacitive sensors. In 2002, Matsumoto released a highly publicized report on spoofing fingerprint sensors using gelatin-based materials [9]. His team used 11 commercially available sensors, both optical and capacitive, and achieved a level of success that pushed spoofing issues to the forefront of research challenges. Readers interested in exactly how to create fake fingers are encouraged to read Matsumoto's publication. Researchers at Clarkson University demonstrated a technique for spoofing sensors using fingerprints placed on Play-Doh and gelatin molds with a success rate of almost 90%. They devised a liveness detection algorithm that uses perspiration patterns around pores on live skin. Using this liveness detection algorithm, the incorrect detection of fakes reduced from 90% to 10%.

There are several well-known antispoofing methods for fingerprint recognition. The most common is measuring the electrical properties of the skin. The optical transmission properties of skin at various wavelengths are different, which can be used to detect the liveness of the source. Multispectral fingerprint sensors use this method for antispoofing. Previously discussed anatomical measurements of perspiration patterns and elastic distortion of the finger skin can provide useful information for detecting spoofing attacks. Ongoing attempts at spoofing fingerprint sensors continue to be published, and these reports are necessary for strengthening the overall security of fingerprint recognition systems.

15.2.2 Face Recognition

Most commercial face recognition systems use a 2-D image for feature extraction and matching. The most common method of spoofing face recognition systems is to present a high-resolution printout or photograph of the individual to be impersonated to the camera. This attack has been carried out successfully against commercially available systems [6]. The same research group has also used a recorded video of genuine users and presented that as sample input. Antispoofing methods that test the liveness of eyes have been implemented in face recognition, but these can be deceived by cutting out holes in the eyes of the image and using it as a mask on a real person. Range images that provide depth information can also be used to determine if a real face or a photograph is being presented to the camera. The face is capable of presenting several expressions with minimal effort and a challenge response system can be designed that directs the users to blink their eyes, turn their faces, and produce expressions such as a frown or smile. The human skin exhibits different optical properties such as reflection and scattering under illumination of varying wavelengths that can be measured for liveness detection [10].

15.2.3 Iris Recognition

Iris recognition systems use an infrared illuminated image of the iris for feature extraction and matching. A very simple attack used a high-quality printout of an iris image and presented it to the iris recognition system. Iris recognition systems that test the liveness of the eye by measuring the dilation of the pupil can be circumvented by cutting out the pupil and using the iris image as mask on a real eye. A 2-D Fourier spectrum analysis of an iris image captured from a live eye and a printed iris has a different distribution, as shown in Figure 15.2 [11]. The Computer Electronic Security Group (CESG) of the U.K. government has conducted laboratory experiments with cosmetic contact lenses. Such lenses have different patterns printed on them that hide the true iris pattern of the user. To test its effectiveness in deceiving iris systems, a user enrolled his eye wearing such a cosmetic lens. After successful enrollment, the same contact lens was placed in a different eye and was aligned to match the radial orientation of the enrollment. The iris system accepted this verification attempt, which provided evidence that such an attack is possible. Precisely such an attack was tested successfully by printing an iris pattern on a contact lens [12].

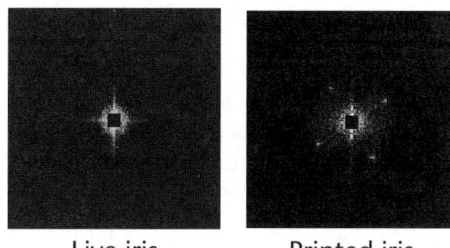

Live iris   Printed iris

Figure 15.2 Fourier analysis of images from a live iris and a printed iris. (© 2003 World Scientific Publishing Co. [11].)

In another experiment researchers printed iris images from 27 individuals on six different paper types after applying four different preprocessing techniques. Two different attack scenarios were created. In attack 1 printed iris images were used for enrollment and verification, and in attack 2 a live iris image was used for enrollment and a printed iris image was used for verification. At an FAR of 0.1% and an FRR of 12.71%, success rates of 57.41% and 49.32% were observed on attack 1 and attack 2, respectively [13].

Several liveness detection techniques have been researched to counteract spoofing attacks. Printed iris images cannot change their geometric proportions if the pupil contracts or dilates. The pupil size is affected by the impinging infrared light, which can be monitored to ensure liveness. Hemoglobin in blood has specific absorption properties under infrared illumination. The vascular structure in the sclera can be used to detect if the eye being presented is live. Iris systems can use challenge response procedures that require users to blink their eyes or look in a certain direction thereby, determining liveness of the source. Currently, a gap exists between existing liveness detection methods and the implementation of these methods in commercial systems, which will need to be addressed to satisfy the privacy and security concerns of the end users.

15.2.4 Hand Recognition

Attacks on hand recognition systems have shown that they can be spoofed using fake hands in a laboratory setting. Chen et al. demonstrated two different methods for spoofing hand recognition systems [14]. The first method used a plaster to recreate an exact mold of the user's hand. The second method used silhouette images captured by the hand recognition system to recreate a cardboard cutout of the hand. Both of these attacks were successful in deceiving the hand recognition systems. Liveness detection methods that assess the biological properties of the hand can be used to mitigate such attacks. The surface temperature of the hand is different compared to the surface temperatures of other materials. A challenge response process that asks the user to lift a particular finger when it is placed on the platen could also be effective.

15.2.5 Speaker Recognition

Mimicry can deceive humans into believing the identity of a voice source, but this technique does not work with voice recognition systems. Instead, a replay attack in which a recorded voice is played back to the system is a more realistic threat. Spoofing speaker recognition systems using replay attacks has been studied since the 1970s. Text-prompted systems, which are based on the challenge response paradigm, are an effective means of countering recorded voice replay attacks. Speaker recognition has applications in criminalistics and forensics, and individuals may try to disguise their voices to avoid detection. A research study examining the abilities of four different classification rules was tested for its ability to pick up a voice that had been disguised by placing a hand over the mouth, pinching a nostril, and speaking at a high pitch and a low pitch [15]. The results showed promise and work is ongoing in this area.

15.2.6 Vascular Pattern Recognition

Future of Identity in the Information Society (FIDIS), a consortium of academia and industry organizations that is part of an EU research program, conducted spoofing attacks on back-of-the-hand vascular pattern systems as part of a larger report on the use of biometrics for identity management [16].The back-of-the-hand vein pattern was captured using a camcorder under infrared illumination. The vein pattern was printed on paper, wrapped on a live hand, and successfully presented to the capture sensor. The particular sensor came equipped with liveness detection technology, but the details are not publicly available. Spoofing techniques that use paper or other substrates with biometric traits printed on them can be detected using Fourier analysis. Biological characteristics of the skin also provide a degree of certainty of source liveness and should be incorporated into these systems. Vascular pattern recognition is still a relatively new technology and is likely to undergo several enhancements as its adoption increases.

Table 15.2 lists a summary of liveness indicators that can be used in various biometric technologies.

15.2.7 Mimicry Attacks

The previous attacks were based on physiological biometric technologies. Technologies that use behavioral biometrics, such as dynamic signature, keystroke dynamics, and gait, cannot be spoofed using physical artifacts. Such systems can be deceived

Table 15.2 Summary of Liveness Indicators

Modality	Physical Properties	Biological Indictors
Fingerprint	Elasticity, distortion, electrical properties (capacitance), reflection, and absorption (skin)	Pulse, temperature and thermograms, perspiration pattern
Face	Reflection and absorption (skin), depth mapping, and color and texture analysis	Temperature and thermograms
Iris	Reflection and absorption (muscle)	Change in pupil size, eyeball movement
Hand	Electrical properties (capacitance), reflection, absorption, depth mapping, and color and texture analysis	Pulse, temperature, and thermograms
Vascular pattern	Reflection and absorption (blood), depth mapping, and color and texture analysis	Pulse, temperature, and thermograms

by mimicry attacks, where an individual impersonates the behavior of another individual. The most common form of mimicry that everyone is likely to have experienced is based on voice. Dynamic signature verification is relatively difficult to mimic because of the inherent behavioral traits, and there were no published studies at the time of this writing. Gaufor et al. conducted a mimicry attack on gait recognition based on information collected by three accelerometers connected to the hips of users [17]. Their results showed that similarity scores of imposters increased when they consciously attempted to copy another person's walking style. Liveness checking measures cannot detect mimicry attacks because they are performed by humans. Rather, detection methods should focus on the technical performance of differentiating between genuine users and imposters, as well as innovative challenge response processes involving secret knowledge known only to the genuine user.

15.2.8 LiveDet Competition

Several of the spoofing attacks described in the previous section require cooperative subjects, which is not a practical situation. Some of these attacks can be carried out by coercion as well, but all types of security systems are susceptible to those. Thus, while some of these attacks might not be carried out on real-world systems, the ability of biometric systems to distinguish a live sample from a fake sample needs to be addressed in future systems. Liveness detection is imperative for antispoofing techniques to be effective. The past few years have seen an increase in interest in software-based liveness detection techniques. In 2009 the first open competition, Fingerprint Liveness Detection Competition, was organized with the aim of assessing the current capabilities in fingerprint liveness detection [18]. A genuine and fake fingerprint database was created using three commercially available sensors. An open call for participation allowed university research teams, industry vendors, and independent researchers to take part in the competition. Of all the algorithms evaluated, the lowest overall error rate for misclassification was 14.6% and the highest error rate was 25.0%. The dataset used in the competition is also available for download by contacting the organizers. This competition has laid the foundation for such activities to be organized for other biometric technologies as well. At the time of this writing, LiveDet II, a follow up to the original competition, was being administered and results from this competition were expected towards the end of 2011 [19]. LiveDet II has introduced a new section for system-based spoof detection where an entire system comprising a sensor and software will be provided by the participants, thereby allowing them to use hardware-based techniques along with software-based methods.

15.2.9 TABULA RASA

The TABULA RASA project initiated under the European Commission ICT Project in FP7 will address the issues of direct spoofing attacks on trusted biometric systems [20]. The TABULA RASA project has the following objectives:

- Address the gap in standards for antispoofing techniques;
- Propose countermeasures to antispoofing techniques;

- Examine new biometric technologies that are inherently robust to direct attacks.

At the time of this writing, the Competition on Counter Measures to 2D Facial Spoofing Attacks was being conducted to assess the performance of algorithms in detecting printed photo attacks on face recognition systems [21]. The results were expected to be released by the end of 2011.

15.3 Standards

Security techniques, specifically those related to the protection of biometric systems and biometric data, are being addressed by the standards community. The SC 27 "Security Techniques" group within ISO/IEC JTC 1 creates standards for generic methods and techniques for IT security. SC 27, in liaison with SC 37, has initiated and published several standards in an attempt to address the gaps that exist in security standardization of biometric systems and data. ISO/IEC 19792:2009 *Security Evaluation of Biometrics* is a standard that specifies the areas that need to be addressed during a security evaluation of a biometric system [22]. The standard establishes a framework for security testing and the evaluation of biometric algorithms and components under varying environmental, application, and privacy requirements. The framework is generic and has three components: testing biometric performance, defining assurance levels, and evaluating the security process. ISO/IEC 24761:2009 *Authentication Context for Biometrics (AcBio)* defines the data structure and data elements by which a service provider can judge the results of a biometric processing unit [23]. There are six processing units defined in this standard: capture, intermediate feature extraction, final feature extraction, storage, matching, and decision. Each of these units uses PKI to sign the data generated by them. The ISO/IEC 24745 *Biometric Template Protection* project, currently being drafted, addresses specific requirements of biometric information privacy such as unlinkability, irreversibility, confidentiality, and data minimization. Technical Committee 68 Financial Services has published ISO 19092:2008, which is a multipart standard that focuses on the secure management of biometric systems in the financial services sector [24]. This standard defines requirements for managing and securing biometric information throughout the information life cycle as it is transmitted between different systems. The integrity protection of biometric data and results generated by the biometric systems are addressed in this standard. At the time of this writing, a new project ISO/IEC NP 30107, titled "Anti-Spoofing and Liveness Detection Techniques," was in the drafting stages. The aim of this proposed standard is to specify the methodologies for preventing attacks on biometric sensors [25].

15.4 Synthetic Biometric Samples

Synthetic biometric samples are simulated biometric samples generated using computer models of the data source. Synthetic biometrics has been mainly explored from a testing and evaluation perspective. Collecting real biometric data for system evaluations is a costly and time-consuming activity and also has some inherent privacy risks for subjects participating in the evaluation. Synthetic biometric methods can be used to generate samples that represent operational conditions that may be difficult to reproduce in the lab environment. Research in synthetic biometrics has examined various methods for generating biometric samples that are modeled on variables such as subject demographics, environmental conditions, and sensor type. These samples are not representative of a particular individual; rather, these samples represent a subgroup that shares certain traits.

Along with performance evaluation, synthetic biometrics can also be used for exploiting system vulnerabilities. Sensor spoofing attacks can be initiated by creating a synthetic biometric sample. A digital representation of the synthetic biometric can be used in transmission attacks and hill-climbing attacks. The use of synthetic biometric samples in exploiting security vulnerabilities has not yet received much attention, but is likely to as synthetic biometric techniques become more sophisticated. Synthetic generators for fingerprints, face, iris, voice, and dynamic signatures already exist and ongoing research is examining other biometric traits.

15.4.1 Synthesis Approaches

There are various methods for synthesizing biometric samples that are biologically similar to human traits. These methods can be generally categorized into physical and statistical modeling techniques [26]. Physical modeling techniques take into consideration the physics of how the underlying trait is created and how it reacts to environmental conditions such as illumination and surface pressure. External influences such as stress and strain have an impact on biometric samples. The impact of these influences can be modeled into the synthesis process. Statistical modeling techniques use empirically derived information about biometric samples to parametrize into a mathematical model that is then used to synthesize samples. This technique requires the analysis of a large amount of data that is used to generate statistical parameters that control the synthesis process. This process has also been called the *analysis by synthesis paradigm* [27].

SFinGe is a synthetic fingerprint generator that is commercially available and was developed at the Biometric System Laboratory at University of Bologna [28]. SFinGe uses shape, a density map, a directional map, and a deformation model created from the empirical evaluation of a large set of fingerprints. SFinGe can also introduce human interaction impact such as rotation and translation in the fingerprints, as shown in Figure 15.3. Another method for synthetic fingerprint generation models the natural formation of ridgelines on the fingerprint [29]. A free demo is available through the Biometric System Laboratory Web site [30].

Synthetic face generation methods use models for various landmarks of the face to create a synthetic face image. Once the attributes about the different face parameters are given to the algorithm, the nose, eyes, lips, and eyebrows are generated using the region-specific models. These are accumulated and fitted to a general face structure to generate a synthetic face image. FaceGen from Singular Inversions, Inc., creates 3-D face images based on the statistical modeling of a human face shape [31]. The software has more than 150 controls including age, illumination, expression, and others that can be used in the synthetic generation of faces.

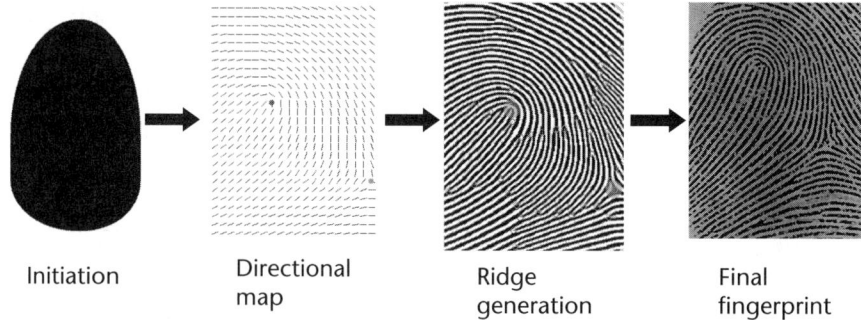

Figure 15.3 SFinGe Synthetic fingerprint based on capacitive sensor.

There are several synthetic iris generation methods. Synthetic iris images can be created by combining several segments of different real iris images and then rotating and translating them. Alternative techniques have superimposed several different iris patterns to create a new iris pattern by combination. Methods based on creating iris patterns based on Fourier transforms of random signals have also been examined.

Readers interested in technical details of synthetic biometrics are encouraged to read [32].

15.5 Summary

Biometric systems face threats from various sources, and these will become more innovative as the technology advances. It is impossible to achieve 100% security, but system designers should take into account some of the more widely exploited vulnerabilities and incorporate safeguards against them. The use of multimodal biometric systems does not eliminate spoofing attacks, but increases the level of effort of such attacks. Readers familiar with information security concepts will have recognized the overlap of biometric system security issues with other types of enterprise IT systems, but spoofing is specific to biometrics and requires special attention, both from the research community as well as vendors and implementers. A lot of commercially available systems today come equipped with antispoofing methods, but these countermeasures increase the processing time and sometimes reduce the performance by increasing the FTE rate, resulting in real-world implementations favoring switching off the antispoofing module in return for faster processing and increased performance. A holistic approach involving basic research and best practices is required to tackle this issue. This chapter has provided an overview of general biometric system vulnerabilities along with a detailed discussion of spoofing methods with the aim of increasing awareness among practitioners, developers, and implementers. With proper security controls biometric systems can provide identity assurance with the level of trust that is required of today's systems.

References

[1] Ratha, N. K., R. M. Bolle, and J. H. Connell, "Enhancing Security and Privacy in Biometrics-Based Authentication Systems," *IBM Systems Journal*, Vol. 40, 2001.

[2] Bartlow, N., and B. Cukic, "Threats and Countermeasures," *Proceedings of Biometrics Consortium Conference*, Washington, D.C., 2005.

[3] Soutar, C., "Security Considerations for the Implementation of Biometric Systems," *Automatic Fingerprint Recognition Systems*, 2004, pp. 415–431.

[4] Galbally, J., et al., "An Evaluation of Direct Attacks Using Fake Fingers Generated from ISO Templates," *Pattern Recognition Letters*, Vol. 31, 2010, pp. 725–732.

[5] Cole, S. A., *Suspect Identities: A History of Fingerprinting and Criminal Identification*, Cambridge, MA: Harvard University Press, 2001.

[6] Thalheim, L., J. Krissler, and P. -M. Ziegler, *Body Check: Biometric Access Protection Devices and Their Programs Put to the Test*, 2002.

[7] Schuckers, S., "Spoofing and Anti-Spoofing Measures," *Information Security Technical Report*, Vol. 7, 2002, pp. 56–62.

[8] Van Der Putte, T., and J. Keuning, "Biometric Fingerprint Recognition: Don't Get Your Fingers Burned," *IFIP TC8/WG8.8 Fourth Working Conference on Smart Card Research and Advanced Applications*, 2000, pp. 289–303.

[9] Matsumoto, T., et al., "Impact of Artificial 'Gummy' Fingers on Fingerprint Systems," *Proceedings of SPIE Optical Security and Counterfeit Deterrence Techniques IV*, 2002.

[10] Parziale, G., J. Dittmann, and M. Tistarelli, *Analysis and Evaluation of Alternatives and Advanced Solutions for System Elements*, BioSecure, 2005.

[11] Daugman, J., "Demodulation by Complex-Valued Wavelets for Stochastic Pattern Recognition," *International Journal of Wavelets, Multi-Resolution and Information Processing*, Vol. 1, 2003, pp. 1–17.

[12] von Seelen, U. C., "Countermeasures Against Iris Spoofing with Contact Lenses," *Biometrics Consortium Conference*, Arlington, VA, 2005.

[13] Ruiz-Albacete, V., et al., "Direct Attacks Using Fake Images in Iris Verification," *Biometrics and Identity Management*, B. Schouten et al., (eds.), Berlin, Germany: Springer-Verlag, 2008, pp. 181–190.

[14] Chen, H., et al., "Fake Hands: Spoofing Hand Geometry Systems," *Biometrics Consortium Conference*, Arlington, VA, 2005.

[15] Perrot, P., and G. Chollet, "A Question of Disguised Voice," *Acoustics 2008*, Paris, France, 2008.

[16] WP6, *D6.1 Forensic Implications of Identity Management Systems*, 2006.

[17] Gafurov, D., E. Snekkenes, and P. Bours, "Spoof Attacks on Gait Authentication System," *IEEE Transactions on Information Forensics and Security*, Vol. 2, September 2007, pp. 491–502.

[18] Marcialis, G. L., *First International Fingerprint Liveness Detection 2009*, 2009, http://prag.diee.unica.it/LivDet09.

[19] LiveDet, "LiveDet II Fingerprint Liveness Detection Competition 2011," 2011, http://people.clarkson.edu/projects/biosal/fingerprint/index.php.

[20] "Trusted Biometrics Under Spoofing Attacks," *TABULA RASA*, 2011, http://www.tabular-asa-euproject.org/project.

[21] "IJCB 2011 Competition on Counter Measures to 2D Facial Spoofing Attacks," *TABULA RASA*, 2011, http://www.tabularasa-euproject.org/evaluations/ijbc-2011-competition-on-counter-measures-to-2d-facial-spoofing-attacks.

[22] ISO/IEC, *ISO/IEC 19792:2009 Security Evaluation of Biometrics*, Geneva, Switzerland, 2009.

[23] ISO/IEC, *ISO/IEC 24761:2009 Authentication Context for Biometrics*, Geneva, Switzerland, 2009.

[24] ISO/IEC, *ISO 19092:2008 Biometrics—Security Framework*, Geneva, Switzerland, 2008.

[25] ISO/IEC, *ISO/IEC NP 30107 Anti-Spoofing and Liveness Detection Techniques*, ISO/IEC JTC 1 SC 37, 2011.

[26] Buettner, D. J., "Biometric Sample Synthesis," in *Encyclopedia of Biometrics, Vol. 2*, S. Z. Li and A. K. Jain, (eds.), Berlin, Germany: Springer, 2009, pp. 116–122.

[27] Yanushkevich, S., et al., "Introduction to Synthesis in Biometrics," in *Image Pattern Recognition: Synthesis and Analysis in Biometrics*, S. N. Yanushkevich, (ed.), New York: World Scientific Publishers, 2007, pp. 6–25.

[28] Cappelli, R., "SFinGe: An Approach to Synthetic Fingerprint Generation," *International Workshop on Biometric Technologies (BT2004)*, Calgary, Canada, 2004, pp. 147–154.

[29] Kuecken, M. U., and A. C. Newell, "A Model for Fingerprint Formation," *Europhysics Letters*, Vol. 68, 2004, pp. 141–146.

[30] "SFinGe," *Biometric System Laboratory, University of Bologna*, http://biolab.csr.unibo.it/ResearchPages/sfinge_demo_download.html.

[31] "FaceGen—3D Human Faces," Singular Inversions Inc., http://www.facegen.com/index.htm.

[32] Yanushkevich, S. N., "Synthetic Biometrics: A Survey," *International Joint Conference on Neural Networks*, 2006, pp. 676–683.

CHAPTER 16

Privacy Concerns in Biometric Applications

Privacy, in its philosophical sense, is generally accepted as a universal right of individuals, but in a practical sense very difficult to enforce. Citizens in countries with privacy laws generally expect a certain level of protection from intrusion and information gathering [1]. The advent of e-governance initiatives and the digitization of citizen data have brought a lot of attention to the protection of privacy, and rightfully so. Although the idea of privacy is largely influenced by geopolitical and cultural factors, the thought of losing control over personal and sensitive information is generally unsettling. The advantage of biometric technologies arises from a strong binding of a user with physiological or behavioral traits. When biometric technologies are placed within an ecosystem that is becoming increasingly digitized and networked, concerns about privacy are bound to arise. Several governments have created privacy protection frameworks to secure personal information collected from citizens. The convergence of biometrics and privacy has received a lot of attention in the recent past and will continue to do with the increasing adoption of large-scale biometric systems. It is inevitable that privacy concerns with respect to biometric technologies need to be examined and addressed. In order to do so, the conceptual foundations of privacy need to be mapped to the technical foundations of biometrics. This chapter will discuss the various biometric technology factors that have raised privacy and cultural concerns, existing legal frameworks to enhance privacy, the role of biometrics in enhancing privacy, and efforts of the standards community in addressing privacy concerns.

16.1 Privacy Invasive Technologies (PIT)

The use of biometric technologies raises privacy concerns because of their potential to violate the principles of observability, linkability, and anonymity of an individual. Privacy advocates fear that biometric technologies can violate the privacy of an individual, and this section will discuss the reasons that are most often raised for the categorization of biometrics as PIT. In addition, readers interested in the sociocultural impact of biometrics in military applications are encouraged to read [2].

16.1.1 Permanent Identifier

The very basic property of biometrics that makes it a strong authentication technology also gives rise to a privacy concern that is voiced most often: it is permanent and thus cannot be revoked. Once a digital identity is bound to a particular biometric trait, it is now a permanent part of the system. The general notion that this identifier cannot be revoked or modified, which is only partially correct, is still a vulnerability which can be exploited. Critics often use the analogy that biometrics are like passwords that cannot be changed, and, unlike passwords, humans only have a limited number of fingers, face, eyes, and other biological traits.

16.1.2 Function Creep

Function creep is a very real threat in networked systems that store large amounts of personal information. Function creep occurs when the original scope of information use is widened with or without the knowledge of the person providing this information. Function creep can happen due to a number of reasons: to improve operational efficiency, to assist law enforcement, or to support the mission of other agencies without duplicating efforts. Data sharing is known to happen between various governments agencies, and the use of biometric identifiers to correlate previously unlinkable information without the user's consent is a privacy threat.

16.1.3 Tracking and Profiling

The use of large-scale identity systems with centralized data storage has raised concerns about anonymity and linkability. Although tracking already happens in today's world using other means such as credit card and debit card transactions, the ability to track an individual or his or her actions using a permanent identifier does not offer any recourse in maintaining anonymity. Biometric systems that operate covertly are even a bigger threat to privacy because of their tracking ability. Sir Francis Galton, who made several important discoveries in biometrics and laid the scientific foundation of fingerprint recognition, was originally attracted to biometrics due to his interest in eugenics and criminalistics. The concept of predicting the membership of a certain subset of the population based on biometric traits is generally considered unethical, but the mere capability of doing so poses a risk.

16.1.4 Data Breach

Previous reports of data breach in government and private enterprise has lead to severe repercussions, both in terms of credibility and financial liability. Data breach of personally identifiable information (PII) such as biometric data can lead to a public backlash, irrespective of the actual impact of the data breach.

16.1.5 Data Misuse

Identity fraud and the misuse of personal data have become more of an issue with the increasing use of e-commerce, e-governance, and other electronic services. Biometric technologies are much harder to spoof and defraud, especially in a supervised system. However, spoofing attacks have been conducted successfully on biometric sensors, which are possible on sensors that operate from a remote and unsupervised location. Certain medical conditions can be deduced from the biometric data that can be of a highly personal and sensitive nature. Information about age, gender, and ethnicity can be extracted from samples used for face recognition. Arthritis and other musculo-skeletal disorders can be potentially detected using fingerprint and hand recognition. The amount of medical information that can be deduced from raw biometric samples and processed templates is more of a theoretical conjecture, but that is indicative of research that is necessary in this area.

16.1.6 Unauthorized Collection

The collection of biometric data without the user's explicit knowledge or consent is possible using covert sensors. The use of CCTV technology definitely makes this a real threat to privacy. The use of multimodal biometric systems also increases the potential of capturing more data than that which is being disclosed to the users.

16.2 Privacy-Enhancing Technologies (PET)

A privacy-enhancing technology (PET) balances the requirements of the protection of personal information, such as the prevention of unnecessary or unauthorized disclosure or processing, with the functionality of the system [3]. Biometric technologies inherently do not embody any PIT or PET characteristics; it is their implementation and operational policies that push them in one direction or another. There is a dedicated field of research that is attempting to eliminate vulnerabilities in biometric systems, data processing, and associated policies and bring biometrics closer to being a PET. Section 16.1 provided an overview of biometric privacy concerns, and all of them revolved around protection and processing of biometric data. The majority of privacy enhancements of biometrics have focused on template protection by performing matching operations in the cryptographic domain, and a summary of the various techniques is given in Table 16.1. Readers interested in a detailed discussion of the various techniques are encouraged to read [4].

The Trusted Revocable Biometric Identities (TURBINE) project, initiated as a research project under the European Union Seventh Framework Programme (FP7) for Research and Technology Development, represents the most focused

Table 16.1 Summary of Biometric Template Protection Techniques

Technique	Description
Salting	Apply transformation to the biometric sample using a secret key. Transformation is reversible and the sample can be retrieved using a secret key.
Noninvertable transformation	Apply a one-way function to a biometric sample that is a nonreversible operation.
Key-binding system	Biometric sample is fused with a cryptographic key and stored as a template.
Key-generating system	Cryptographic key is extracted from the biometric sample.

Source: [4].

international effort of designing and demonstrating practical biometric encryption techniques. TURBINE has created a technical framework that should fulfill the following requirements:

- Template protection;
- Renewable and revocable templates;
- Applicable to any biometric technology and architecture;
- Interoperable;
- Support verification using a minimal amount of information.

The technical foundations of the framework rest on the notions of *pseudo identities* (PI) and *auxiliary data* (AD). A PI is generated for each application accessed by a user and is used in subsequent verification operations. The AD allows a user to generate multiple, independent PIs that are unlinkable. The TURBINE framework stores the PI as the enrollment template and uses it for matching purposes. The AD can be either stored on a centralized database or kept in the possession of the user. Readers interested in a detailed discussion of the TURBINE framework are encouraged to read [5].

The main technical challenge in any type of a biometric cryptosystem arises from the conflicting properties that repeated biometric samples from the same individual will never be exactly the same, and cryptographic techniques require every data bit to be the same for a successful operation. Readers interested in the specific technical challenges of biometric cryptosystems should read [6].

16.3 Privacy Frameworks

Governments are required to collect PII from its citizens for the efficient administration of its services, and citizens have a right to be assured that their personal information will be kept confidential. Since the 1960s governments have paid increasing attention to protection of data collected from its citizens and formulated privacy regulations that can be enacted and enforced. The earliest citizen data protection regulations were implemented in Hessen in Germany in 1970, followed by Sweden in 1973, France in 1978, and other countries [7]. The Organization for Economic Co-operation and Development (OECD) provides a forum for its member countries to discuss, develop, and refine economic and social policies. In 1980 the OECD released *Guidelines on the Protection of Privacy and Transborder Flows of Personal Data*, which laid out principles to take into account when formulating and implementing policies for data protection and data flow [8]. The following OECD guidelines, also called Fair Information Practices, have formed the basis for privacy regulations of several countries around the world:

1. The principle of limited collection;
2. The principle of data quality;
3. The principle of scope definition;
4. The principle of limited usage;
5. The principle of security safeguards;

6. The principle of openness;
7. The principle of individual participation;
8. The principle of accountability.

In 1995 the European Union (EU) released the Directive 95/46/EC, also known as the Data Protection Directive, with the specific goal of protecting privacy of individuals by setting guidelines for processing of personal data by automatic or manual methods [9]. This directive does not set forth specific processes and procedures, but does provide a framework of the rights and obligations to which the member states of the EU should adhere. These include transparency, clear limitation of purpose, maintaining the data quality, the subject's right to information, special protection of sensitive data, confidentiality of processing, creation of oversight authority, method for the remedy of data, and transfer of data to other countries. Although there is agreement on the general principles progressed by the directives, the implementation of the directive has not been uniform across member states because of differences in the interpretation of specifics. Biometrics is not explicitly mentioned in this directive; rather, the interpretation of personal data is kept open to accommodate advances in technology and the most significant impact is on interpretation of biometric data as personal information. There is a general consensus that raw biometric samples should be treated as personal information, but differences arise with respect to processed biometric data. Certain member states treat biometrics data as personal data only if it can be linked to an individual. In case the biometric data is encrypted using a private key, it is not considered to be personal data in its encrypted format. Other member states such as Belgium, Denmark, and France have taken the view that, irrespective of the biometric data being encrypted, it should be treated as personal data. With an increasing flow of data between nations, a consistent interpretation of privacy regulations will be critical.

Privacy and Identity Management in Europe (PRIME), a research project initiated as part of the European Union's Sixth Framework Programme, aims to create a privacy-enhancing identity management system that integrates functional, legal, economic, and social requirements. PRIME has established a list of technical principles that guide its implementation principles [10]:

- Design must start from maximum privacy.
- Explicit privacy rules govern system usage.
- Privacy rules must be enforced.
- Privacy enforcement must be trustworthy.
- Users need an easy and intuitive manifestation of privacy.
- Privacy needs an integrated approach.

PRIME offers technical and architectural requirements that provide detailed information for technologists, as well as a framework with nontechnical components that can be used to guide the design process. Readers interested should read the PRIME Framework document [11].

Another EU effort in addressing the privacy and ethical aspects of biometrics is the Rising Pan European and International Awareness of Biometrics and Security

Ethics (RISE) project, which was started in 2009 [12]. The overall objective of RISE is to create a forum for discussing the ethical, social, and privacy implications of biometrics and create an international interest group to progress these discussions in parallel with technology developments.

The Privacy Act of the USA, passed in 1974, regulates the collection, storage, use, dissemination, and disclosure of personal data collected from its citizens and permanent residents by the various federal agencies and federal contractors. The Privacy Act also established a code of fair information practices to comply with the statutory norms specified for the collection, use, and dissemination of PII. This act also provides for individuals to pursue civil claims against the government in case of violation of the act. The Computer Matching and Privacy Protection Act of 1988 lists procedural requirements for government agencies where computerized matching is used. The Gramm-Leach-Bliley Act (GLBA) of 1999 contains comprehensive privacy statutes for protecting the private information of clients of financial institutions. The GLBA does not specify the technologies that are required to enforce the statutes but instead focuses on the requirements of privacy protection. The Homeland Security Presidential Directive (HSPD) 24, signed by President Bush in 2008, establishes "a framework to ensure that Federal executive departments and agencies (agencies) use mutually compatible methods and procedures in the collection, storage, use, analysis, and sharing of biometric and associated biographic and contextual information of individuals in a lawful and appropriate manner, while respecting their information privacy and other legal rights under United States law" [13]. Although the main objective of this directive is to promote interoperability and support the mission of the various federal agencies, the framework supports balancing the privacy of individuals with these objectives.

The scope and application of privacy protection laws have taken different evolutionary paths in nations around the world and have been largely influenced by their geographical, political, economic, and cultural factors. These laws also influence citizens' views of their governments and the level of skepticism that they harbor about their governments. Data protection is a prime example of the difference in privacy protection laws between the United States and the EU. EU data protection laws prohibit the EU member countries from sending the PII of EU citizens to countries that do not have similarly adequate laws. The transfer of PII between the law enforcement agencies of the EU and the United States is a cumbersome process, and the United States has sought to create a data protection framework to facilitate data sharing based on principles acceptable to both the EU and the United States. At the time of this writing, such a framework was being drafted by the EU parliament. Readers interested in reading more about privacy laws in different countries are encouraged to read [14].

16.4 Privacy Impact Assessment (PIA)

Many technology implementations, including biometrics, have an implication on the privacy of its users. The PIA is a formalized method of risk assessment from a privacy perspective, and it is being increasingly used by authorities responsible for protecting the privacy of users. Information is now considered to be an asset of strategic value, and it needs to be treated accordingly. The PIA needs to fulfill certain criteria [15]:

- Identify key issues;
- Concentrate on process and technology;
- Have representation from all stakeholders;
- Be sufficiently independent.

The Information Comisioners Office (ICO) in the United Kingdom has developed a five-step framework for conducting a PIA. Although this framework is generic, it can be adapted for biometric systems, processes, and data. An overview of the framework is given in Table 16.2. The complete handbook can be found in [16].

The PIA is a qualitative assessment that can be modified to maximize the benefit for the organization conducting it. The PIA is only as effective as the questions that are discussed with the key stakeholders. The following is a list that can be used as a starting point for creating a more extensive questionnaire:

- To what degree do the users give up control of personal information?
- Will the impact of the biometric system be proportional on all users?
- What is the impact of acquisition errors and mismatch errors?
- What kind of continuity plan is in place?
- Has the system scope been analyzed for function creep?
- Does the biometric system respect the religious and cultural sensibilies of the target population?
- What are the repercussions of a data breach to the user as well as the organization?
- Is there an audit plan for the oversight of the operational biometric system?
- Who is the owner of the biometric information: the user or the owner of the biometric system?

Table 16.2 PIA Framework

Phase	Description
Preliminary phase	Conduct preliminary analysis of privacy issues, identify key stakeholders, collect concerns from all relevant groups, create scope of PIA
Preparation phase	Develop plan for holding effective discussions with all key stakeholders and form a cross-functional group with representation from all key stakeholders
Analysis phase	Identify key privacy concerns and design issues, reengineer project to satisfy these concerns, and review solutions with cross-functional group
Documentation phase	Outline the entire process, solutions, and mitigation measures and map the privacy protection and enhancing features to regulations and legislations
Review and audit phase	Conduct periodic reviews of the implementation and suggest any modifications based on upgrades to the system

Source: [16].

- What type of data loss threats exist in each subcomponent of the biometric system?

For a PIA to be effective and deliver real value, it has to be conducted as part of the planning process and not as a standalone activity. A properly conducted PIA can help organizations make privacy a part of its overall strategy instead of approaching it as an afterthought. The International Biometric Group (IBG) initiated the BioPrivacy^{TM} Initiative and developed the BioPrivacy^{TM} Application Impact Framework, which is a technique used for evaluating the impact of biometric systems on the privacy of its users [17]. The application framework, illustrated in Table 16.3, consists of evaluating a biometric system implementation on 10 assessment questions and assigning a subjective privacy risk value based on the response.

16.5 Standards

Until recently, the national and international biometric standards committees have mainly focused on the technical aspects of the technology, although there have been efforts in the past toward addressing privacy concerns as part of larger reports. Standards create a consistent sense of technical scope and objectives, and this approach is necessary to tackle privacy concerns. Several efforts at the national and international levels have tried to address privacy concerns independently or as part of a larger context. The INCITS M1 Ad Hoc Group on Biometrics and E-Authentication (AHGBEA) published a report in 2007 that discussed the strengths and weaknesses of use of biometrics for e-authentication over open networks [18]. Along with technical details, this report outlined a list of best practices to protect the privacy of users when using biometrics for remote electronic authentication. These recommendations were a mixture of technical controls, policies, and procedures to reduce the breach of user privacy.

Working Group 6 of ISO/IEC JTC 1 SC 37 published a technical report titled *Jurisdictional and Societal Considerations for Commercial Applications—Part 1: General Guidance,* which provides guidelines for handling the privacy and social concerns at various stages of a biometric system [19]. This report is aimed at

Table 16.3 BioPrivacy Application Impact Framework

Assessment Criteria	*Response*	
	Lower Risk	*Higher Risk*
User awareness	Overt	Covert
User participation	Voluntary	Mandatory
Recognition process	Verification	Identification
System accessibility	Private	Public
Status of user	Private employee	Citizen/resident
Owner of biometric information	User	System administrator
Data storage	Smartcard	Centralized database
Type of biometric technology	Behavioral	Physiological
Type of data stored	Templates	Raw samples

Source: [17].

addressing design and implementation issues with respect to the social and legal concerns of the use of biometric data and accessibility by the largest part of the population.

SC 27 *IT Security Techniques* of ISO/IEC JTC1 creates standards for generic methods and techniques for IT security. Strong synergies exist in the work conducted by these groups and have resulted in several standards that cut across multiple committees. ISO/IEC 19792:2009 *Security Evaluation of Biometrics* is a standard that specifies the areas that need to be addressed during a security evaluation of a biometric system [20]. ISO/IEC 24761:2009 *Authentication Context for Biometrics (AcBio)* is a standard that defines the structure and the data elements by which a service provider can judge the results of a biometric authentication operation [21]. At the time of this writing, ISO/IEC 24745 *Biometric Template Protection* was being drafted, which addresses the specific requirements of biometric information privacy such as unlinkability, irreversibility, confidentiality, and data minimization using cryptographic techniques. The goal of this standard is to provide a technical framework for improving privacy protection by securing the biometric templates. Biometrics-based PETs are expected to benefit from this work.

Another subcommittee, SC 17 *Cards & Personal Identification*, has worked on standardization projects aimed at the use of biometrics in card-based architectures. ISO/IEC 7816:2004—Part 11 provided guidelines for the use of integrated circuit cards in personal verification using biometric technologies [22]. ISO/IEC 24787:2010 establishes requirements for performing comparisons of biometric samples and returning decisions on an integrated circuit card and for security policies for on-card biometric comparison [23].

ISO 19092:2008, prepared by Technical Committee 68 Financial Services, is a standard specifically for the secure management of biometric systems in the financial services sector [24]. This standard defines requirements for managing and securing biometric information throughout its life cycle and is implicitly linked to addressing the privacy concerns of users. It also includes a tutorial on the formulation of security controls for the protection of biometric systems, "Biometric Policies and Biometric Practice Statements," and it journals events for audit purposes. Although this standard is specific to the financial services sector, it provides guidelines that are transferable to other sectors and also presents a framework for addressing similar issues in other industry sectors.

Industry consortia have also taken up the challenge of drafting privacy practices and informal standards. One such example is the *Biometric Privacy Code* published by the Biometrics Institute in Australia. This code establishes guidelines on how biometric data should be collected, used, and disclosed to protect the privacy of users [25]. The overarching goals of the code are:

1. To facilitate the protection of personal information provided by or held by biometric systems;
2. To facilitate the process of identity authentication in a manner consistent with the Privacy Act;
3. To promote biometrics as a privacy-enhancing technology (PET).

The code is nonbinding; members of the Biometrics Institute pledge to implement it, which in turn creates a trust relationship with the end users.

16.6 Privacy-Enhancing Design Principles

Addressing privacy concerns as an afterthought is no longer possible. Information assets are becoming more valuable and thus require that commensurate efforts are made to protect it from misuse. Privacy goes beyond just data protection and should be addressed accordingly. Privacy enhancement needs to be addressed as a design requirement and not just as a functional requirement by making it an integral part of project planning and organizational policies. Table 16.4 provides a list of privacy-enhancing design principles that should be incorporated into every biometric system design.

16.7 Future Challenges

The basic privacy concerns related to biometrics are discussed in Section 16.1, but as technology advances, more challenges are bound to arise. Cloud computing is a relatively new paradigm that enables convenient, on-demand network access to data and resources that are managed as part of a larger pooled infrastructure. Cloud computing architecture is characterized by sharing resources and the abstraction of data management methods. Cloud systems that manage biometric information need to address several privacy concerns, specifically with regards to data ownership and data sharing.

Behavioral biometric technologies have been explored with relatively less interest than physical biometric technologies. Profiling of individuals, especially by governments, has increased as a result of security threats and so have the privacy concerns of individuals. Current research is examining the predictability of intent based on neurological factors such as ocular changes, thermal changes to skin, respiratory rates, and other related behavioral factors. Although such applications are not a classic application of biometric technologies, the scientific underpinnings of behavioral biometrics can potentially be used for such purposes.

Table 16.4 Privacy-Enhancing Design Principles

Category	Principles
Organizational	Commitment to privacy as an organizational priority and in measurable terms
	Establish best practices and policies for incorporating privacy into operations
	Establish PIA framework and the continual application of it
	Methods for performing periodic audits and reviews
Technical	Design should not require a compromise between technical functionality and privacy requirements
	Use privacy-enhancing technologies (PET)
	Apply security controls to ensure the confidentiality, integrity, and availability of the data
	Compliance with standards and regulations
User	Minimization of data collection
	Use disclosure
	Redress mechanism
	Retention and revocation mechanism

The flow of information between organizations and nations is increasing as they try to improve operational efficiency and security of borders. E-passports with biometric information embedded in them are an example of the flow of sensitive information that needs to be regulated. The lack of internationally recognized standards for guiding organizations in implementing privacy controls is an obstacle to achieving consistent and uniform privacy practices across nations, organizations, and industries. This needs to be addressed by bringing together researchers and government and industry representatives.

Current legislation needs to evolve as the scope of collection and the usage of PII increase. This also raises an interesting dichotomy. Supporters of biometric technologies advance the argument that legislation curtails it from achieving its full potential, whereas its critics claim that legislation only legitimizes its privacy-invasive nature. Assuaging the concerns of these extremes is unlikely, but a mixture of policies, privacy impact assessments, and research into biometrics as a PET is the way forward.

16.8 Summary

The benefits of using biometric systems need to be balanced with the potential privacy concerns that it raises. Data protection and privacy protection frameworks already exist in various forms, but their applicability to biometric systems and data needs a deeper examination. These issues can affect the acceptability of a biometric system irrespective of its other benefits and thus hamper its adoption. Media today plays a prominent role in spreading information about biometric technologies, and some of it is inaccurate. Educating users and increasing awareness can go a long way in dispelling the fears of users, but that is not enough. The role of vendors is also crucial in addressing this challenge. They are the best equipped to build privacy-enhancing techniques such as the encryption of data communication and liveness detection in sensors. A holistic approach consisting of technical solutions, educational programs, and policy is necessary to address privacy concerns, and such a holistic approach needs to be incorporated in the design, development, and deployment of biometric systems.

References

[1] Tavani, H. T., and J. H. Moor, "Privacy Protection, Control of Information, and Privacy-Enhancing Technologies," *Computers and Society*, Vol. 31, 2001, pp. 6–11.

[2] Woodward, J. D., et al., *Army Biometric Applications: Identifying and Addressing Sociocultural Concerns*, RAND Corporation, 2001.

[3] van Blarkom, G. W., J. J. Borking, and J. G. E. Olk, *Handbook of Privacy and Privacy-Enhancing Technologies: The Case of Intelligent Software Agents*, The Hague, the Netherlands: College bescherming persoonsgegevens, 2003.

[4] Jain, A. K., K. Nandakumar, and A. Nagar, "Biometric Template Security," *EURASIP Journal on Advances in Signal Processing*, Vol. 2008, 2008, pp. 1–18.

[5] TURBINE, *D2.3.3—Research Findings for Standardisation*, 2010.

[6] Uludag, U., et al., "Biometric Cryptosystems: Issues and Challenges," *Proceedings of the IEEE*, Vol. 92, June 2004, pp. 948–960.

[7]   Neuwirt, K., "Convention 108: New Challenge for Data Protection in Non-European States," 2008, http://www.infodf.org.mx/web/participantes/neuwirt/Abstract%20/Karel%20Neuwirt.pdf.

[8]   OECD, *Guidelines on the Protection of Privacy and Transborder Flows of Personal Data*, 1980, http://www.oecd.org/document/18/0,3343,en_2649_34255_1815186_1_1_1_1,00.htm.

[9]   EU, "Protection of Individuals with Regard to the Processing of Personal Data and on the Free Movement of Such Data," 1995.

[10] EU, *Privacy and Identity Management in Europe*, 2004.

[11] Fischer-Hübner, S., and H. Hedbom, *PRIME Framework V3*, PRIME, 2008.

[12] RISE, "Rising Pan European and International Awareness of Biometrics and Security Ethics," 2009.

[13] *HSPD 24—Biometrics for Identification and Screening to Enhance National Security*, Washington, D.C., 2008, http://www.biometrics.gov/Documents/NSPD59%20HSPD24.pdf.

[14] NBSP, *International Data Privacy Laws and Application to the Use of Biometrics in the United States*, 2006.

[15] Stewart, B., "Privacy Impact Assessments," *Privacy Law & Policy Reporter*, Vol. 3, 1996, pp. 61–64.

[16] ICO, *Privacy Impact Assessment Handbook Version 2*, U.K. Information Commissioners Office, 2009.

[17] International Biometric Group, "BioPrivacy™ Initiative: Biometrics and Privacy."

[18] INCITS M1.4, *Study Report on Biometrics in E-Authentication*, 2007.

[19] ISO/IEC, *ISO/IEC TR 24714-1:2008 Jurisdictional and Societal Considerations for Commercial Applications—Part 1: General Guidance*, Geneva, Switzerland, 2008.

[20] ISO/IEC, *ISO/IEC 19792:2009 Security Evaluation of Biometrics*, Geneva, Switzerland, 2009.

[21] ISO/IEC, *ISO/IEC 24761:2009 Authentication Context for Biometrics*, Geneva, Switzerland, 2009.

[22] ISO/IEC, *ISO/IEC 7816-11:2004-Integrated Circuit Cards—Part 11: Personal Verification Through Biometric Methods*, Geneva, Switzerland, 2004.

[23] ISO/IEC, *ISO/IEC 24787:2010 Identification Cards—On-Card Biometric Comparison*, Geneva, Switzerland, 2010.

[24] ISO/IEC, *ISO 19092:2008 Biometrics—Security Framework*, Geneva, Switzerland, 2008.

[25] *Biometrics Institute Privacy Code*, Biometrics Institute, Crows Nest, NSW, Australia, 2009.

Acronyms

ANSI	American National Standards Institute
ABIS	Automated Biometric Identification System
AFIS	Automated Fingerprint Identification System
BAT	Biometric Automated Toolset
CBEFF	Common Biometric Exchange Formats Framework
CDEFFS	Committee to Define an Extended Fingerprint Feature Set
CMC	cumulative match characteristic
CODIS	Combined DNA Index System
DET	detection error trade-off
DHS	Department of Homeland Security
DoD	Department of Defense
EBTS	Electronic Biometric Transmission Specification
EFTS	Electronic Fingerprint Transmission Specification
FAR	false accept rate
FERET	facial recognition technology
FMR	false match rate
FNIR	false negative identification rate
FNMR	false nonmatch rate
FPIR	false positive identification rate
FRR	false reject rate
FRVT	Face Recognition Vendor Test

FTA	failure to acquire
FTE	failure to enroll
FVC	Fingerprint Verification Competition
HIIDE	Handheld Interagency Identification Device
IAFIS	Integrated Automatic Fingerprint Identification System
INCITS	International Committee for Information Technology Standards
IREX	Iris Recognition Exchange
ISO/IEC	International Organization for Standardization/International Electrotechnical Committee
JTC	Joint Technical Committee
MBE	Multiple Biometric Evaluation
MBGC	Multiple Biometric Grand Challenge
MINEX	Minutiae Interoperability Exchange
NBIS	NIST Biometric Image Software
NFIQ	NIST Fingerprint Image Quality
NGI	next generation identification
NIST	National Institute of Standards and Technology
PIV	Personal Identity Verification
ROC	receiver operating characteristic
SFinGe	Synthetic Fingerprint Generator
SMT	scar mark tattoo
SRE	speaker recognition evaluation
SVC	Signature Verification Competition
WSQ	wavelet scalar quantization

Glossary

1:1 Comparison   The process of comparing an input biometric sample to only one template. See *Verification*.

1:N Comparison   The process of comparing an input biometric sample to more than one template. See *Identification*.

Acquisition   The process of capturing the raw biometric sample from a user.

Active sensing   In the context of image acquisition, these are technologies that utilize an external illumination source, such as infrared illumination.

Attempt   An acquisition process that results in a calculation of a similarity score.

Anthropometry   The science of measuring physical traits of humans for understanding individual and group variability and classification purposes.

Behavioral biometric technologies   Biometric technologies use actions or mannerisms that are acquired or learned over time such as signature, gait, and typing pattern.

Biological biometric technologies   Biometric technologies use anatomical features such as fingerprints, face, and iris structure.

Biometrics   The automated recognition of humans based on biological or behavioral characteristics.

Binarization   The process of creating an image with only black and white pixels.

Binning   The process of storing biometric samples in classes based on a higher-level relationship in data attributes.

Candidate list   The result of an identification operation that returns a list of the closest matching candidates to the input sample.

Cepstral features   Features produced as a result of performing a Fourier transform on a log spectrum of a speech signal.

Closed-set identification   Determines if an input sample belongs to an individual who is a member of the system and already known to the system.

Core   The point of maximum curvature on the innermost ridgeline.

Crossover error rate See *Equal Error Rate.*

Cumulative match characteristic curve A performance evaluation curve that represents all possible rank values on the x-axis and the probability of correct identification at each possible rank value on the y-axis.

Delta The point where two ridgelines moving in parallel change direction and move away from each other.

Detection error trade-off curve A performance evaluation curve that plots false match and false nonmatch rates on the x-axis and y-axis as a function of the threshold.

Dynamic signature verification Verification process that uses features such as the velocity, direction, number of strokes, time of each stroke, and pressure applied by the user.

Eigenface Eigenvectors that are calculated from the face image covariance matrix and represent the essential characteristics of the face image.

Enrollment The individual provides his or her biometric sample to the system and a template is generated and stored for future use.

Equal error rate The performance point where FAR and FRR are equal.

Failure to acquire rate The probability of user attempts during identification or verification for which the system cannot acquire an appropriate sample.

Failure to enroll rate The proportion of user enrollment transactions that cannot be completed according to the enrollment policy.

False acceptance rate The proportion of verification transactions from imposters that will be incorrectly accepted.

False match rate The proportion of samples from imposter attempts that are successfully matched against the enrolled templates of genuine users.

False nonmatch rate The proportion of samples from genuine attempts that cannot be matched against the enrolled templates of genuine users.

False rejection rate The proportion of verification transactions from genuine users that will be incorrectly rejected.

False positive identification rate The proportion of identification transactions performed by nonenrolled users that returns a candidate list of which they are a member.

False negative identification rate The proportion of identification transactions performed by enrolled users that returns a candidate list of which they are not a member.

Freeman chain code Method used to represent the contour of the hand as an enclosed object.

Function creep   The expansion of the original scope of the use of information with or without the knowledge of the person providing this information.

Gallery   The list of templates that are used for comparison against an input sample in identification.

Galton features   The features formed by discontinuities in the flow of ridges on finger skin, also called *minutiae*.

Gait recognition   The process of recognizing a person based on walking rhythm.

Genuine attempt   A user tries to match his or her sample against his or her own enrollment template.

Habituation   The process of learning how to interact with a biometric system that results in improved performance and a better user experience.

Hamming distance   Algorithm that computes the difference in bits between two vectors as a measure of dissimilarity.

Henry classification   The classification system based on overall ridge flow created by Edward Henry for manual fingerprint identification.

IDENT   Automated biometric identification system managed and operated by the U.S. Customs and Border Protection Agency.

Identification   The process in which the individual does not make any claim to an identity when providing his or her biometric to the system.

Identity management   The process of identifying an individual and controlling access to resources based on his or her associated privileges.

Imposter attempt   A user tries to match his or her sample against another user's enrollment template.

Intraclass variance   Variance in feature representations of the same modality from the same source.

Interclass variance   Variance in feature representations of the same modality from different sources.

Interocular distance   Distance between the center of the two eyes of an individual.

ISO/IEC   International Organization for Standardization/International Electrotechnical Commission, responsible for creating international standards.

IrisCode   A 512-byte iris template created by the algorithm designed by John Daugman.

Latent fingerprint   Fingerprint that is created by deposits of natural oils secreted by the skin that are left on surfaces with which a finger has come in physical contact.

Lights out   Identification process that does not require any human intervention for adjudication purposes.

Live scan systems   Fingerprint sensors that capture a digitized representation of a fingerprint using optical, capacitive, and other technologies.

Liveness detection   The automated process of determining if a sample was acquired from a live source.

Match score   See *Similarity score*.

Match-on-card   Architecture that uses a smart card that is capable of storing and matching biometric samples.

Minutiae   The points formed by discontinuities in the ridgelines of fingerprints.

Multibiometric system   A system that fuses information from multiple biometric traits, algorithms, and sensors and systems.

Near-infrared illumination   Illumination in the wavelength of 700–900 nm.

Negative identification   The process of proving to the system that the user is not known to the system.

Nonrepudiation   The ability of an individual to not deny performing a transaction.

Normalization   The process of converting two different datasets so that they conform to a predefined set of common parameters.

Off-line comparison   Process of creating an enrollment template and matching separate from the sample capture.

Online comparison   Process of creating an enrollment template and matching during the sample capture.

Open-set identification   Determines if an input sample belongs to an individual who is a member of the system but does not have to be enrolled in the system.

Operational evaluation   An evaluation to determine the performance of a complete biometric system deployed in a real-world application that is being used by a particular user population.

Operator   Individual responsible for operating a biometric system or supervising the enrollment, verification, or identification process.

Positive identification   The process of proving to the system that the user is known to the system.

Penetration rate   The proportion of a database that needs to be searched for identification purposes, which leads to a successful match.

Presentation   A user interaction with a sensor that results in a sample being produced for further processing.

Privium system Subscription-based border control application using iris recognition administered at the Schipol Airport in Amsterdam, the Netherlands.

Range image Face image that consists of depth information for every pixel.

Rank The smallest-sized candidate list of which the user is a member.

Receiver operating characteristic (ROC) curve A performance evaluation curve that plots true accept rates versus false accept rates on the x-axis and y-axis, respectively.

Ridge bifurcation The point at which a single ridgeline divides into two or more ridgelines.

Ridge ending The point at which a ridgeline comes to an abrupt end.

Scenario evaluation An evaluation to determine the overall performance of a biometric system in a simulated application environment that is representative of the real-world application environment.

Score distribution The distributions of genuine and imposter match scores.

Segmentation The process of separating the region of interest from the entire sample representation.

SFinGe Synthetic fingerprint generator available for research and commercial purposes through the Biometric System Laboratory at the University of Bologna, Italy.

Similarity score The degree of confidence that the two feature representations being compared are from the same individual.

Spectrogram A time series graph of frequency or amplitude of a sound signal.

Spoofing The act, or process, of presenting a nonlive or fake biometric sample.

Synthetic biometrics Simulated biometric samples generated using computer models of the data source.

Technology evaluation An evaluation conducted to assess the performance of a biometric system for a target application or to compare multiple systems against a common sample database.

Template The feature representation that is stored as a reference for future recognition operations.

Test crew Set of subjects who will participate in a biometric system evaluation.

Threshold The cutoff value for the variability allowed in two feature representations from the same individual.

Throughput The number of users that a system can correctly process in a given amount of time.

Token A tangible object that an authorized user possesses to prove his or her identity.

Transaction The completion of one or more attempts for the purpose of enrollment, verification, or identification.

Utility The observed or predicted positive or negative contribution of a biometric sample to overall performance of a biometric system.

Verification The process in which an individual makes a claim to an identity when providing his or her biometric to the system.

Watch list A list of individuals of interest that is used in negative identification applications.

Zero effort imposter attempt An attempt in which an individual presents his or her own biometric sample for verification against his or her own template, but the comparison is made against another individual's template.

About the Author

Shimon Modi is a technology specialist on biometric applications with more than 8 years of experience in applied research and standards development. Dr. Modi was the director of research at the Biometric Standards, Performance, and Assurance Laboratory at Purdue University, where he also taught a graduate-level course on applied biometrics. Dr. Modi has served as a visiting biometric scientist at the Center for Development for Advanced Computing (C-DAC), an R&D agency for the Government of India, where he helped establish the National Biometric Lab. Dr. Modi's interests reside in the application of biometrics to e-authentication, the statistical analysis of system performance, and enterprise-level information security. He has also been involved in U.S. and international biometric standards development in the past. Dr. Modi has written and contributed to four books and more than a dozen scientific articles and is a frequent presenter at conferences.

Index

A

Acquisition, 7
Authentication Context for Biometrics (AcBio), 182
AFIS, 46
ANSI, 48, 172, 173
Application
    Types, 12
    Classification, 13
Attempt, 17
    Genuine attempt, 18
    Imposter attempt, 18
Automated fingerprint identification system, 46
Automated speech recognition, 107
Auxiliary data, 240

B

Behavioral biometrics, 4
BioAPI, 163, 174, 177, 181, 217
    Architecture, 178
BioAPI consortium, 174
Biometrics,
    Architecture, 217
    Definition, 1, 2
    Encryption, 240
    Template, 7
    Types, 4
    System model, 7
Biometric Applications,
    Government, 12
    Commercial, 12
    Forensic, 13
    Personalization, 13
Biometric Automated Toolset, 91
BioVisa, 73
British Standards Institute, 173

C

Candidate list, 9
Common Biometric Exchange Format Framework (CBEFF), 176
Cepstrum, 108
Character, *see* Sample quality
CODIS, 150
Comparative Biometric Testing, 204
    Round 6, 128
    Round 7, 146, 200
Conformance testing, 181
Core, 41
Cumulative match characteristic curve, 22, 24

D

Database,
    BioSec, 139
    BiosecurID, 139
    BioSecure Multimodal, 139
    Dynamic signature, 202
    Face recognition, 196
    Fingerprint recognition, 191
    Iris recognition, 198
    MOBIO, *see* Evaluation
    Multibiometrics, 203
    Quality in Face and Iris Research Ensemble (Q-FIRE), 203
    Speaker recognition, 201
    Speech sample databases, 118
    UBIRIS, 197
Data storage, 7
Decision making, 8
Deformable models,
    3-D face recognition, 66
Delta, 41
Dermatoglyphics, 33
Detection error tradeoff curve, 23

Disaster recovery, 219
DNA recognition, 149
Doddington's zoo, *see* Zoo analysis
Dynamic signature verification,
    Digitizer pads, 135
    Feature based representation, 136
    Function based representation, 136
    Offline, 134
    Online, 134
    Universal background model, 137
Dynamic time warping, 113, 136

E

Electronic Biometric Transmission Specification,
    DoD EBTS, 48, 183
    FBI EBTS, 47, 183
Edward Henry, *see* Henry Classification
Eigenfaces, 62
Elastic bunch graph matching (EBGM), 63
Encryption, 226, 240
Enrollment, 8
Equal error rate, 21
Errors
    Acquisition errors, 18
    Generalized, 21
    Matching errors, 19
    Type I, 23
    Type II, 23
Evaluation,
    FERET, 68, 191
    FpVTE, 48, 189
    FRGC, 70, 193
    FRVT, 69, 192
    Fingerprint Verification Competition (FVC), 50, 189
    ICE, 87, 196
    IREX-I, 89, 197
    IREX-II, 90
    IRIS06, 88 197
    ITIRT, 86, 195
    Liveness detection, 205, 231
    MBE, 70, 194
    MBGC, 70, 193
    MINEX, 49, 189
    MOBIO, 71, 116
    Noisy Iris Challenge Evaluation (NICE), 197
    Operational evaluation, 28
    Scenario evaluation, 28
    Signature verification competition, 138, 200
    Technology evaluation, 28

F

Face recognition,
    2-D, 60
    3-D, 65
    2-D acquisition, 60
    2-D segmentation, 60
    2-D normalization, 61
    Appearance based matching, 62
    Feature based matching, 62
    Image quality, 61
Face Verification Competition, 71
Failure to enroll rate, 19
Failure to acquire rate, 19
False accept rate, 20
False match rate, 20
False negative identification rate, 23
False nonmatch rate, 20
False positive identification rate, 23
False reject rate, 20
FERET, *see* Evaluation
Fidelity, *see* Sample quality
Finger geometry, 98
Fingerprint capture,
    Electro-optical sensor, 39
    Inked capture, 36
    Live scan, 36
    Multispectral sensor, 39
    Optical sensor, 37
    RF sensor, 38
    Solid state capacitive sensor, 37
    Thermal sensor, 38
Fingerprint feature extraction, 42
Fingerprint matching,
    Correlation based matching, 45
    Minutiae based matching, 44
    Ridge based matching, 45
Fingerprint Verification Competition (FVC), *see* Evaluation
Fist of sender, *see* Keystroke dynamics
FpVTE, *see* Evaluation
Francis Galton, 34
FRGC, *see* Evaluation

Freeman Chain Code, 100
FRVT, *see* Evaluation
Fusion, *see* Multibiometric fusion

G

Gabor filter, 63, 83
Gait recognition, 150
Gaussian Mixture Model (GMM),
    Dynamic signature verification, 137
    Hand recognition, 101
    Speaker recognition, 113
Genuine attempt, *see* Attempt
Generalized false match rate, 21
Generalized false reject rate, 21

H

Habituation, 89, 129, 213
Hamming distance, 84
Hand geometry,
    Image acquisition, 98
    Feature extraction, 99
    Feature matching, 101
    Pegs, 98
Hemoglobin, 94, 123, 229
Henry classification, 35
Hidden Markov model, 114
Hill climbing, 223-227
Hough transform, 45, 83

I

Integrated Automated Fingerprint Identification
    System (IAFIS), 52
Identity management, 1, 12, 165, 210
Iris Challenge Evaluation (ICE), *see Evaluation*
Identification, 9
    Closed set, 10
    Negative, 10
    Open set, 10
    Positive, 10
    Rank, 22
Imposter attempt, *see* Attempt
Infrared, 5, 60, 65, 70, 86, 94, 98, 123, 195
Integrated automated fingerprint identification
    system, 52
IREX, *see* Evaluation
Iris,
    Acquisition, 80

Ciliary zone, 79
Collarette, 79
Crypt, 79
Extraction, 82
Image quality, 83
IrisCode, 80, 83, 84
IRIS06, *see* Evaluation
ISO, 172
ISO/ IEC JTC 1 SC 37,
    Working group 1, 175
    Working group 2, 176
    Working group 3, 177
    Working group 4, 178
    Working group 5, 179
    Working group 6, 179
ITIRT, *see* Evaluation
ITU-T, 175

J

JPEG2000,
    Fingerprint, 44, 49
    Face, 67, 75
    Iris, 85
    Vein pattern, 128

K

Keystroke dynamics, 143
    Features, 144

L

Likelihood ratio, 113, 160
Liveness detection, 3,

M

Matching, 7
MBE, *see* Evaluation
MBGC, *see* Evaluation
MFCC, *see* Speaker recognition
MINEX, *see* Evaluation
Minutiae point, 33
Multibiometrics, 6
    Levels of fusion, 156
    Score normalization, 159
    System types, 154
Multibiometric fusion,
    Decision level, 161
    Feature level, 157
    Rank level, 161

Multibiometric fusion (continued)
    Score level, 157
    Sensor level, 156
    Quality based, 162

N

Negative Identification, see Identification
NFIQ, 43
NICE, 90
NIST, 174
Nonrepudiation, 1

O

OASIS, 174
Ocular recognition, 200
Operational evaluation, see Evaluation
Optical path, 113

P

Performance, 3
Permanence, 3, 33
Physiological biometrics, 4
Personal Identity Verification (PIV), 53
Polar coordinates, 83, 85, 89, 90
Policy development, 218
Positive identification, see Identification
Presentation, 17
PRIME, 241
Principal Component Analysis (PCA), 60, 62, 101,
Privacy, 216
    Data breach, 238
    Data misuse, 238
    Function creep, 238
    OECD, 240
    Permanent identifier, 238
    Tracking and profiling, 238
Privacy Enhancing Technology, 239
Privacy Impact Assessment, 242
Privacy Invasive Technology, 237
Privium, 15, 91
Psudeo identity, 240

Q

Quality, see Sample quality

R

Rank, see Identification

Range image, 66
Recognition,
    DNA recognition, 149
    Dynamic signature, 133
    Face recognition, 59
    Fingerprint recognition, 33
    Gait recognition, 150
    Hand recognition, 97
    Iris recognition, 79
    Keystroke dynamics, 143
    Retina recognition, 148
    Speaker recognition, 107
    Vein pattern recognition, 123
Retina recognition, 148
Revocable, 240
RISE, 241
Rolled fingerprints, 40

S

Sample quality,
    Character, 30
    Fidelity, 30
    Utility, 30
Scenario evaluation, see Evaluation
Score normalization, see Multibiometrics
Seafarers' Identity Document, 11
Security analysis, 223
SFinGe, 233
Short tandem repeat, 149
Signal processing, 7
Silhouette, 99
Similarity score, 7
Slap fingerprints, 40
Speaker recognition,
    Information levels, 110
    MFCC, 112
    Microphone effect, 120
    Text-dependent, 109
    Text-independent, 109
    Text-prompted, 109
    Universal background model, 113
    Windowing, 111
Speaker recognition evaluation, 115, 198
Spectrogram, 108
Spoofing, 3, 227
    Fingerprint, 227
    Face, 228

Iris, 228
　　Hand, 229
　　Speaker, 229
　　Vascular pattern, 230
Standards, 171
　　Sample quality, 180
　　Working Group, 172
Standards Developing Organization, 172
Standoff iris system, 85
Stereo imaging system, 65
Structure light system, 65
Support vector machines (SVM), 114, 145
Synthetic biometrics, 233
System integration, 211
　　Framework, 221

T

TABULA RASA, 231
Technology evaluation, *see* Evaluation
Template, 7
Template adaptation, 101
Threshold, 8
Throughput, 3
Token image, 67
Transaction, 17
TURBINE, 239

U

UAE expellee program, 91
Usability, 213
Usability design, 215

User specific analysis, *see* Zoo analysis
Utility, *see* Sample quality
US-VISIT, 52

V

Variance,
　　Interclass variance, 8
　　Intraclass variance, 8
Vascular pattern recognition, *see* Vein recognition
Vein recognition, 123
　　Reflective capture, 124
　　Transmissive capture, 124
Verification, 8
Video surveillance, 64
Visible spectrum, 64
Voice recognition, *see* Speaker recognition
Vulnerabilities,
　　Process, 226
System, 224
　　Transmission, 225

W

Watch list, 73, 224
Wavelet Scalar Quantization, 43
Working groups, *see* Standards

Z

Zero effort attempt, 141
Zoo analysis, 25

Recent Titles in the Artech House Information Security and Privacy Series

Rolf Oppliger, Series Editor

Biometrics in Identity Management: Concepts to Applications, Shimon K. Modi

Bluetooth Security, Christian Gehrmann, Joakim Persson and Ben Smeets

Computer Forensics and Privacy, Michael A. Caloyannides

Computer and Intrusion Forensics, George Mohay, et al.

Defense and Detection Strategies against Internet Worms, Jose Nazario

Demystifying the IPsec Puzzle, Sheila Frankel

Developing Secure Distributed Systems with CORBA, Ulrich Lang and Rudolf Schreiner

Electric Payment Systems for E-Commerce, Second Edition, Donal O'Mahony, Michael Peirce, and Hitesh Tewari

Evaluating Agile Software Development: Methods for Your Organization, Alan S. Koch

Homeland Security Threats, Countermeasures, and Privacy Issues, Giorgio Franceschetti and Marina Grossi, editors

Identity Management: Concepts, Technologies, and Systems, Elisa Bertino and Kenji Takahashi

Implementing Electronic Card Payment Systems, Cristian Radu

Implementing the ISO/IEC 27001 Information Security Management System Standard, Edward Humphreys

Implementing Security for ATM Networks, Thomas Tarman and Edward Witzke

Information Hiding Techniques for Steganography and Digital Watermarking, Stefan Katzenbeisser and Fabien A. P. Petitcolas, editors

Internet and Intranet Security, Second Edition, Rolf Oppliger

Introduction to Identity-Based Encryption, Luther Martin

Java Card for E-Payment Applications, Vesna Hassler, Martin Manninger, Mikail Gordeev, and Christoph Müller

Multicast and Group Security, Thomas Hardjono and Lakshminath R. Dondeti

Non-repudiation in Electronic Commerce, Jianying Zhou

Outsourcing Information Security, C. Warren Axelrod

Privacy Protection and Computer Forensics, Second Edition,
   Michael A. Caloyannides

Role-Based Access Control, Second Edition, David F. Ferraiolo, D. Richard Kuhn, and
   Ramaswamy Chandramouli

Secure Messaging with PGP and S/MIME, Rolf Oppliger

Security Fundamentals for E-Commerce, Vesna Hassler

Security Technologies for the World Wide Web, Second Edition, Rolf Oppliger

SSL and TLS: Theory and Practice, Rolf Oppliger

Techniques and Applications of Digital Watermarking and Content Protection,
   Michael Arnold, Martin Schmucker, and Stephen D. Wolthusen

User's Guide to Cryptography and Standards, Alexander W. Dent and
   Chris J. Mitchell

For further information on these and other Artech House titles, including previously considered out-of-print books now available through our In-Print-Forever® (IPF®) program, contact:

Artech House
685 Canton Street
Norwood, MA 02062
Phone: 781-769-9750
Fax: 781-769-6334
e-mail: artech@artechhouse.com

Artech House
16 Sussex Street
London SW1V 4RW UK
Phone: +44 (0)20 7596-8750
Fax: +44 (0)20 7630-0166
e-mail: artech-uk@artechhouse.com

Find us on the World Wide Web at: www.artechhouse.com